Jörg Michel Anderlohr

Modélisation de la combustion dans la ligne d'échappement d'un moteur

Jörg Michel Anderlohr

Modélisation de la combustion dans la ligne d'échappement d'un moteur

Postoxydation des gaz émis d'un moteur à allumage commandé

Presses Académiques Francophones

Impressum / Mentions légales
Bibliografische Information der Deutschen Nationalbibliothek: Die Deutsche Nationalbibliothek verzeichnet diese Publikation in der Deutschen Nationalbibliografie; detaillierte bibliografische Daten sind im Internet über http://dnb.d-nb.de abrufbar.
Alle in diesem Buch genannten Marken und Produktnamen unterliegen warenzeichen-, marken- oder patentrechtlichem Schutz bzw. sind Warenzeichen oder eingetragene Warenzeichen der jeweiligen Inhaber. Die Wiedergabe von Marken, Produktnamen, Gebrauchsnamen, Handelsnamen, Warenbezeichnungen u.s.w. in diesem Werk berechtigt auch ohne besondere Kennzeichnung nicht zu der Annahme, dass solche Namen im Sinne der Warenzeichen- und Markenschutzgesetzgebung als frei zu betrachten wären und daher von jedermann benutzt werden dürften.

Information bibliographique publiée par la Deutsche Nationalbibliothek: La Deutsche Nationalbibliothek inscrit cette publication à la Deutsche Nationalbibliografie; des données bibliographiques détaillées sont disponibles sur internet à l'adresse http://dnb.d-nb.de.
Toutes marques et noms de produits mentionnés dans ce livre demeurent sous la protection des marques, des marques déposées et des brevets, et sont des marques ou des marques déposées de leurs détenteurs respectifs. L'utilisation des marques, noms de produits, noms communs, noms commerciaux, descriptions de produits, etc, même sans qu'ils soient mentionnés de façon particulière dans ce livre ne signifie en aucune façon que ces noms peuvent être utilisés sans restriction à l'égard de la législation pour la protection des marques et des marques déposées et pourraient donc être utilisés par quiconque.

Coverbild / Photo de couverture: www.ingimage.com

Verlag / Editeur:
Presses Académiques Francophones
ist ein Imprint der / est une marque déposée de
AV Akademikerverlag GmbH & Co. KG
Heinrich-Böcking-Str. 6-8, 66121 Saarbrücken, Deutschland / Allemagne
Email: info@presses-academiques.com

Herstellung: siehe letzte Seite /
Impression: voir la dernière page
ISBN: 978-3-8381-7496-9

à ma mère

Table des matières

Nomenclature

Symbole	Signification	Unités[1]
A	Facteur pré exponentiel / Pre-exponential Factor	$[s^{-1}]$
$ABDC$	Après le point mort bas / After Bottom Dead Center	[-]
AI	Auto-inflammation / Auto-Ignition	[-]
ALE	Arbitrary Lagrangian Eulerian	[-]
$ATOL$	Tolérance absolue / Absolute TOLerance	[-]
$RTOL$	Tolérance relative / Relative TOLerance	[-]
AV / CAD	Angle Vilebrequin / Crank Angle Degree	$[°]$
c	Variable d'avancement / Progress variable	[-]
CAI	Auto-inflammation controlée / Controled Auto Ignition	[-]
CFD	Computational Fluid Dynamics	[-]
CFR	Cooperative Fuel Research Engine	[-]
C_p	Capacité calorifique spécifique à pression constante / Specific heat capacity at constant pressure	$[J\ mol^{-1}\ K^{-1}]$
C_v	Capacité calorifique spécifique à volume constant / Specific heat capacity at constant volume	$[J\ mol^{-1} K^{-1}]$
CPU	Central Processor Unit	[-]
CR	Taux de compression / Compression Ratio	[-]
$Diff$	Diffusion / Diffusion	[-]
e	Energie spécifique / Specific energy	$[J\ kg^{-1}]$
E_a	Energie d'activation / Activation Energy	$[J\ mol^{-1}]$
EGR	Recirculation des gaz brûlés / Exhaust Gas Recirculation	[-]
$ECFM3Z$	3-Zones Extended Coherent Flame Model	[-]
ETBE	Ethyl-Tert-butyl-éther / Ethyl-Tert-butyl-ether	[-]
FLR	Réacteur tubulaire / Flow Reactor	[-]
FPI	Flame Prolongation of ILDM (Intrinsic Low Dimensional Manifold)	[-]
h	Enthalpie spécifique / Specific Enthalpy	$[kJ\ g^{-1}]$
h_P	Constante de Planck / Planck constant	$[J\ s]$
$HCCI$	Homogeneous Charged Compression Ignition	[-]
IAE / SAI	Injection d'Air à l'Echappement / Secondary Air Injection (SAI)	[-]
JSR	Réacteur auto-agité / Jet Stirred Reactor	[-]

---

[1] Les unités utilisées correspondent au système SI, sauf indiquées différemment dans le texte.

I	Entrée / Inlet	[-]
$ILDM$	Intrinsic Low Dimensional Manifold	[-]
k_B	Constante de Boltzmann / Boltzmann constant	[J K^{-1}]
$l_{R\text{-}R'}$	Longeur de la liaison R-R' / Length of the R-R' bond	[m]
L_v	Chaleur latente d'évaporation / Latent heat of vaporization	[kJ mol^{-1}]
M	Masse molaire / Molar mass	[g mol^{-1}]
$\dot{M}$	Débit massique / Mass stream	[kg s^{-1}]
n	Nombre de moles / Number of mols	[mol]
$\Delta n_{int.rot}$	Change in the number of internal rotations as the reactant moves to transition state	[-]
NTC	Coefficient négatif de température / Negative Temperature Coefficient	[-]
Out	Sortie / Outlet	[-]
PRF	Carburant de référence primaire / Primary Reference Fuel	[-]
ppm	Particules par Million / Parts per Million	[-]
$ppmc$	Particules Carbonées par Million / Carbon Parts per Million	[-]
p	Facteur stérique (empirique) / Empirically estimated steric factor	[-]
P	Pression / Pressure	[atm]
PSR	Réacteur parfaitement agité / Perfectly Stirred Reactor	[-]
R	Constante des gaz parfait / Perfect gas constant	[J mol^{-1} K^{-1}]
rpd	Nombre d'atomes d'hydrogène transférables/ Reaction Path Degeneracy: Number of abstractalbe H-atoms	[-]
rpm	rotations par minute / revolutions per minute	[min^{-1}]
RON	Indice d'octane de recherche / Research Octane Number	[-]
s	Entropie spécifique / Specific Entropy	[J m^{-3} K^{-1}]
S	Coefficient stoechiométrique / Stoichiometric coefficient	[-]
Sym	Symétrie / Symmetry	[-]
t	temps / time	[s]
T	Température / Temperature	[K]
UHC	Hydrocarbures imbrûlés / Unburned Hydrocarbons	[-]
V	Volume	[m^3]
W	Masse moléculaire moyenne / Mean molecular weight	[g/mol]
$Wall$	Parois / Wall	[-]
y	Fraction massique / Mass fraction	[-]
x	Fraction molaire / Mole fraction	[-]

Lettres grecques

Symbole	Signification	Unités
α	Fraction d'air secondaire / Fraction of secondary air	[kg kg^{-1}]
ϕ	Richesse / Equivalence Ratio	[-]
μ	Masse molaire reduite / Reduced molar mass	[-]
$\dot{\omega}$	Taux de réaction / Reaction rate	[kg m^{-3}s^{-1}]

Indices et exposants

Symbole	Signification	Unités
0	Etat initial / Initial state	[-]
akt	Activation / Activation	[-]
BG	Gaz Brûlés / Burned Gases	[-]
burned	Brûlé / burned	[-]
u	Imbrûlé / Unburned	[-]
comp	calculé / computed	[-]
D	Dilution / Dilution	[-]
eq	Etat d'équilibre / Equilibrium state	[-]
f	Formation / Formation	[-]
FPI	FPI-tabulated characteristic	[-]
H-abstr	Arrachage d'un atome H / H-abstraction	[-]
init	initial / initial	[-]
rec spec	Espèce reconstruite / Reconstructed species	[-]
res	Résiduel / Residual	[-]
T	Traceur / Tracer	[-]
x	Nombre d'atomes de carbone / Number of carbon atoms	[-]
y	Nombre d'atomes d'hydrogène / Number of hydrogen atoms	[-]

La numérotation des réactions est indiquée entre parenthèses	( )
La numérotation des équations est indiquée entre crochets	[ ]

Chapitre 1

Introduction générale

1.1 Les émissions des moteurs à combustion interne

Aujourd'hui, les moteurs à combustion interne avec des carburants essence ou diesel, représentent la principale source d'énergie pour les véhicules automobiles. Ils jouent également un rôle important dans des systèmes stationnaires comme les générateurs d'électricité. Grâce à la très haute densité énergétique des hydrocarbures, les moteurs à combustion interne sont encore difficilement remplaçables en tant que sources d'énergie pour notre société. Néanmoins, la combustion des hydrocarbures est aussi une source de pollution qui peut être la cause de graves problèmes environnementaux et de santé. En conséquence, des restrictions de plus en plus drastiques sur les émissions polluantes des véhicules automobiles sont imposées par la législation. Désormais, la réduction des émissions polluantes provenant des moteurs constitue une source de préoccupation majeure pour les constructeurs automobiles.

Les polluants chimiques majeurs qui sont émis lors de la combustion des hydrocarbures sont les hydrocarbures imbrûlés (UHC), le monoxyde de carbone (CO) et les oxydes d'azote (NO_x). De graves problèmes environnementaux (effet de serre) sont attribués également au dioxyde de carbone (CO_2). Les processus de formation de ces polluants sont décrits ci-après.

Hydrocarbures imbrûlés (UHC) :

Les UHC sont générés, soit par une combustion incomplète, notamment dans le cas d'une combustion en mélange riche, soit par l'émission du carburant condensé sur les parois et dans les crevasses de la chambre de combustion (Heywood, 1988). Les UHC sont constitués principalement du carburant initial, d'alcanes, d'oléfines, d'alcools, d'aldéhydes, de cétones et de composés aromatiques. Piperel et al. (2007) ont mesuré la composition des UHC dans une ligne EGR (Exhaust Gas Recirculation = ligne de recirculation des gaz brûlés) d'un moteur HCCI (Homogeneous Charge Compression Ignition). Ils ont constaté que lors d'une combustion en mélange pauvre ($\phi = 0.7$) avec un fort taux d'EGR (50-65%), entre 30 et 99% des UHC sont composés de carburant initial. Les autres composants majeurs sont produits par l'oxydation partielle du carburant. Ils sont constitués principalement de méthane (CH_4), d'éthylène (C_2H_4) et de formaldéhyde (HCHO). Les UHC font partie des composés cancérogènes et leur émission est la principale source de pollution atmosphérique dans les agglomérations urbaines avec une forte densité de véhicules.

Le monoxyde de carbone (CO) :

Le CO est formé essentiellement lors d'une combustion en mélange riche. Par manque d'oxygène, le carburant n'est pas complètement oxydé en dioxyde de carbone (CO_2) et une large partie est partiellement oxydée en CO. Selon le mode de combustion, la teneur en CO dans les gaz brûlés varie entre des valeurs presque nulles en mélange pauvre (Piperel et al. 2007) et des concentrations supérieures à 5% lors d'une combustion en mélange riche (Errico et al. 2002; Onorati et al. 2003). Pour une température de l'ordre de 1700 K et pendant la phase de détente dans le moteur, les émissions de CO issues de la combustion dans un moteur correspondent à l'équilibre de la réaction (1) (Heywood, 1988),

$$(1) \quad CO_2 + H_2 \leftrightarrow CO + H_2O$$

Le monoxyde de carbone est un gaz toxique. Comme dans le cas des UHC, le CO contribue largement à la pollution atmosphérique des grandes agglomérations.

Les oxydes d'azote (NO_x) :

La production des NO_x est principalement issue des réactions de décomposition de l'azote (N_2) (Onorati et al. 2003). Les réactions de formation du monoxyde d'azote (NO) sont décrites par le mécanisme de Zeldovich (1946) qui a été élargi par Lavoie et al. (1970). Les principales voies de formation de NO sont :

$$(2) \quad \bullet O \bullet + N_2 \leftrightarrow NO + \bullet N \bullet$$

$$(3) \quad \bullet N \bullet + O_2 \leftrightarrow NO + \bullet O \bullet$$

$$(4) \quad \bullet N \bullet + \bullet OH \leftrightarrow NO + \bullet H$$

Due à la forte énergie de liaison entre les atomes d'azote dans les molécules de N_2, ces réactions sont uniquement importantes à haute température ($T \sim 2000$ K). Les températures dans la chambre de combustion atteignent très vite de tels niveaux et les concentrations de NO formées dans le moteur excèdent facilement des niveaux de 1000 ppm dans certains cas. Les NO_x peuvent causer des maladies respiratoires et leur émission contribue à la formation des pluies acides. Deux stratégies permettent la réduction des émissions des polluants.

1. En minimisant leur formation dans la chambre de combustion. Ceci peut être envisagé grâce à des stratégies de combustion optimisées, dont celles utilisées dans les moteurs HCCI ou CAI (Controled Autoignition).

2. En les détruisant dans la ligne d'échappement grâce à des moyens dits « secondaires ». Dans le cas des moteurs à allumage commandé, le catalyseur 3-voies représente un tel moyen.

1.2 Réduction des émissions de polluants à l'aide de la postoxydation

Plusieurs stratégies innovatrices de combustion en moteur, comme celles des moteurs HCCI ou CAI, sont basées sur le principe de la re-circulation des gaz brûlés (EGR). Les moteurs HCCI et CAI combinent certains avantages des moteurs à allumage commandé et des moteurs à allumage par compression. Un mélange gazeux homogène assure une formation réduite de particules et le taux de dilution élevé garantit de faibles émissions de NO_x, ceci grâce à des températures de combustion plus basses. Dans ce type de moteurs, l'auto-inflammation (AI) est contrôlée par le taux et la composition de l'EGR [2] (Lü et al. 2005a , 2005b).

La Figure 1-1 montre le schéma d'un système EGR où un piquage du conduit d'échappement permet de diriger une partie des gaz brûlés vers le conduit d'admission. Le débit de gaz brûlés recirculés est contrôlé par une vanne EGR.

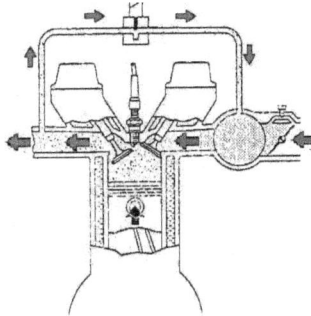

Figure 1-1   Schéma d'un système EGR (Motorlexikon, 2009).

A fort taux d'EGR, l'auto-inflammation est gouvernée par l'injection des gaz brûlés et le milieu réactif dans le moteur est caractérisé par des taux de dilution élévés.

Le catalyseur 3-voies représente une autre stratégie de réduction des émissions de polluants. Aujourd'hui, le catalyseur 3-voies constitue un moyen privilégié pour réduire les émissions d'hydrocarbures imbrûlés, de CO et d'oxydes d'azote (NO_x) formées dans la chambre de combustion d'un moteur à allumage commandé à richesse unitaire (stœchiométrie). Pour fonctionner, le catalyseur doit atteindre une température de 500 K minimum. Pour des températures inférieures, le catalyseur est inefficace et les polluants sont

---

[2] Le taux d'EGR y_{EGR} peut etre défini par le débit des gaz frais $\dot{M}_{frais}$ et le débit d'EGR $\dot{M}_{EGR}$ comme $y_{EGR} = \dfrac{\dot{M}_{EGR}}{\dot{M}_{EGR} + \dot{M}_{frais}}$

émis dans l'atmosphère sans être convertis en produits de combustion totale (idéalement CO_2 et H_2O). Pour cette raison, une grande quantité des polluants est émise pendant le démarrage à froid du moteur. 80 % des émissions d'hydrocarbures sont produites pendant les 40 premières secondes de fonctionnement d'un moteur à allumage commandé (Onorati et al. 2003). Il y a donc un fort intérêt à réduire le temps d'amorçage du catalyseur afin d'atteindre les niveaux d'émissions imposés par une législation de plus en plus stricte. L'injection d'air à l'échappement (IAE) dans un moteur à allumage commandé est un moyen efficace pour accélérer le chauffage du catalyseur lors du démarrage à froid. Le principe de l'IAE repose sur l'injection d'air secondaire derrière la soupape d'échappement lors du passage des gaz brûlés. La Figure 1-2 illustre le principe d'un système d'IAE.

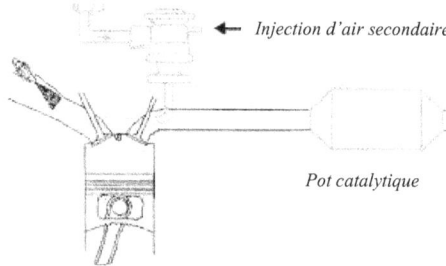

Figure 1-2   Schéma d'un système d'IAE (d'après Onorati et al. 2003).

L'oxygène supplémentaire provenant de l'IAE se mélange avec les UHC issus de la chambre de combustion et peut, dans certaines conditions, s'auto-enflammer. La postoxydation résultante permet d'une part de réduire directement les émissions d'hydrocarbures et de CO, et d'autre part d'augmenter la température des gaz d'échappement, ce qui facilite l'amorçage du catalyseur. *Dans le cadre de ce travail, la postoxydation est définie comme l'oxydation des hydrocarbures dans les gaz brûlés présents dans la ligne d'échappement du moteur.*

1.3   Caractéristiques thermochimiques de la postoxydation dans une ligne d'échappement

Les gaz brûlés sont expulsés brusquement de la chambre de combustion. Ceci crée un écoulement gazeux très pulsé à la sortie des cylindres. Dans le cas de l'IAE, l'air secondaire est injecté derrière la soupape d'échappement. Ceci crée une turbulence supplémentaire dans l'écoulement et des fluctuations de la pression des gaz émis dans le collecteur d'échappement. La pression dans la ligne d'échappement d'un moteur varie autour de la pression atmosphérique (application sans turbo-compresseur). Quand la soupape d'émission s'ouvre, la ligne d'échappement est connectée avec la chambre de combustion. La pression à l'intérieur de la chambre de combustion est supérieure à la pression atmosphérique et de ce fait, lors de l'ouverture de la soupape, des ondes de pression acoustiques sont créées. La Figure 1-3 montre l'évolution de la pression dans la chambre de combustion lors de l'ouverture de la soupape d'échappement d'un moteur à allumage commandé équipé d'un système d'IAE en fonction de l'angle vilebrequin. Les mesures ont été effectuées sur les bancs moteurs de l'IFP (Kleemann 2006). Le moteur testé a été équipé d'un système d'IAE et a fonctionné à une richesse de 1.25.

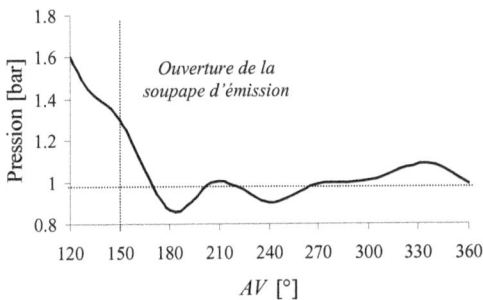

Figure 1-3   *Evolution de la pression dans la chambre de combustion lors de l'ouverture de la soupape d'échappement d'un moteur à allumage commandé équipé d'un système d'IAE en fonction de l'angle vilebrequin (AV) (Kleemann, 2006).*

Lors de la phase de détente, avant l'ouverture de la soupape d'émission ($AV < 150°$), on observe une diminution constante de la pression à l'intérieure de la chambre de combustion. Au moment de l'ouverture de la soupape d'échappement, la dynamique des gaz conduit à une chute de pression vers des pressions inférieures à la pression atmosphérique. Des ondes de choc acoustiques sont produites dans la ligne d'échappement et par réflexion, renvoyées dans la chambre de combustion. Ces ondes acoustiques sont amorties et, à partir d'un angle vilebrequin d'environ 270°, seules des petites fluctuations acoustiques sont détectées.

Kleemann et al. (2006) ont également mesuré l'évolution de la température dans la ligne d'échappement d'un moteur à allumage commandé équipé d'un système d'IAE. Ils ont effectué des mesures pour deux points de la ligne d'échappement, le premier situé dans la culasse du moteur à proximité de la soupape d'échappement, et le second à la sortie du collecteur, devant le pot catalytique. La Figure 1-4 montre l'évolution des températures mesurées lors d'un cycle moteur en fonction de l'angle vilebrequin.

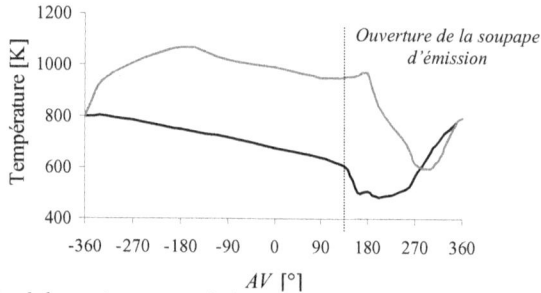

Figure 1-4   *Evolution de la température mesurée dans les gaz brûlés émis par un moteur à allumage commandé équipé d'un système d'IAE (Kleemann, 2006). La ligne noir (▬) correspond aux températures mesurées à proximité de la soupape d'échappement, la ligne rouge (▬) aux températures mesurées à proximité du pot catalytique.*

L'ouverture de la soupape d'émission (~150° d'angle vilebrequin) provoque une chute de la température mesurée dans la culasse du moteur. Ceci est dû à l'air froid injecté par le système d'IAE. Après ~50 ° d'angle vilebrequin, les gaz d'échappement réagissent avec l'air frais et la température mesurée augmente jusqu'à un maximum atteint à ~360° d'angle vilebrequin. Le maximum dépassé, la température diminue constamment avant que le cycle moteur suivant commence. Quant à l'évolution des températures mesurées à proximité du pot catalytique, les températures minimales et maximales sont décalées d'une phase de ~180° d'angle vilebrequin par rapport aux températures mesurées dans la culasse. Ceci correspond au temps de séjour des gaz brûlés dans le collecteur entre la chambre de combustion et le pot catalytique. Grâce à la chaleur générée par la postoxydation, les températures mesurées à proximité du pot catalytique excèdent celles mesurées à la sortie de la chambre de combustion d'environ 200 à 300 K.

Les températures présentées sur la Figure 1-4 correspondent principalement à un régime d'oxydation à basse température. Comparée à une combustion à haute température ($T > 1000$ K), la combustion à basse température obéit à une cinétique chimique plus complexe. Dans un tel régime thermochimique, on observe le phénomène appelé « Coefficient Négatif de température » (en anglais, NTC pour "Negative Temperature Coefficient") qui est caractérisé par une diminution de la réactivité chimique lorsque la température augmente. Ce phénomène, typique de l'oxydation de certains hydrocarbures de la famille des alcanes, est largement décrit dans la littérature (Walker et Morley, 1997 ; Battin-Leclerc, 2008). Pour ce qui est de la composition des gaz brûlés, elle varie fortement en fonction du régime et de la charge, ainsi que des divers réglages du moteur.

En général, les gaz brûlés sont composés :

- de polluants formés dans la chambre de combustion : le CO, les hydrocarbures imbrûlés et les NO_x,

- de produits majeurs de combustion : CO_2, H_2O,

- de N_2 provenant de l'air.

Kleemann et al. (2006) ont mesuré également l'évolution de la concentration en hydrocarbures près de la soupape d'échappement. La Figure 1-5 montre les concentrations d'hydrocarbures mesurées en fonction de l'angle vilebrequin en molécules carbonés par million (ppmc).

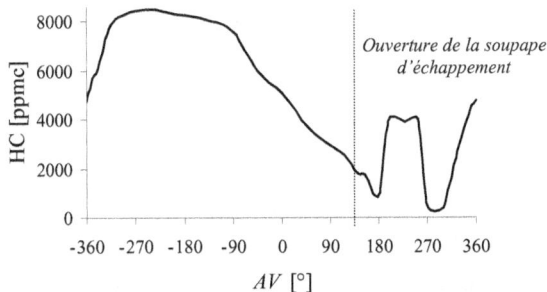

Figure 1-5   Evolution des hydrocarbures imbrûlés mesurés dans la ligne d'échappement d'un moteur à allumage commandé équipé par un système d'IAE (Kleemann, 2006).

Après l'ouverture de la soupape d'échappement (~150° d'angle vilebrequin), les concentrations en hydrocarbures mesurées augmentent de manière drastique. A partir d'environ 200° d'angle vilebrequin, on observe une stagnation de ces concentrations, suivie d'une forte chute observée autour de 270° d'angle vilebrequin. Vers 300° d'angle vilebrequin, la concentration d'hydrocarbures recommence à augmenter jusqu'à un maximum de 8000 ppmc atteinte autour de -270° d'angle vilebrequin. Après la fermeture de la soupape d'échappement, la concentration en hydrocarbures diminue constamment jusqu'à ce que la soupape s'ouvre à nouveau et qu'un nouveau cycle commence. La chute des concentrations d'hydrocarbures observée autour de 270° d'angle vilebrequin est provoquée par l'air frais injecté de manière pulsée par le système d'IAE à proximité de la soupape d'échappement. Ceci éloigne les gaz brûlés des capteurs d'hydrocarbures (blow down) et par conséquent, temporairement, moins d'hydrocarbures sont mesurés. On constate que les hydrocarbures présents dans les gaz brûlés peuvent atteindre des niveaux de concentration de 8000 ppmc.

Les produits majeurs de la combustion, CO_2, H_2O et N_2, dominent largement la composition des gaz brûlés. Ces espèces sont en général considérées comme inertes et donc comme des diluants. Dans le cas d'une combustion quasi-complète, les concentrations des diluants CO_2, H_2O et N_2 dans les gaz brûlés peuvent varier entre des valeurs de 90 % à 100 %. Dans les conditions thermochimiques étudiées durant ce travail, la présence des polluants, tels que le CO ou les NO_x, même en faibles concentrations, peut avoir un effet important sur la réactivité des hydrocarbures. Des changements importants de la cinétique d'oxydation des hydrocarbures en présence des NO_x ont déjà été démontrés (Dubreuil et al. 2006; Glaude et al. 2005; Frassoldati et al. 2003b). D'autres espèces chimiques comme le CO, le H_2 ou les aldéhydes auraient également un effet non négligeable sur le processus d'auto-inflammation de la charge (Subramanian et al. 2007). La maîtrise des phénomènes de postoxydation nécessite une compréhension aussi bien des effets cinétiques, que des effets physiques de mélange et de turbulence.

1.4   Objectifs et plan

Ce travail a comme but le développement d'un modèle numérique prédictif pour la simulation des phénomènes de postoxydation. Un tel modèle devra reproduire le processus d'auto-inflammation des hydrocarbures durant la postoxydation, mais également la décomposition des polluants. Ceci nécessite dans un premier temps de mettre au point un schéma cinétique détaillé qui tienne compte de la chimie à basse température des hydrocarbures et de l'influence sur cette chimie des différentes espèces majeures présentes dans les gaz brûlés. Ces espèces sont CO_2, H_2O et N_2, qui agissent comme des diluants, mais également les polluants tels que CO ou les NO_x, qui même en faibles concentrations, peuvent avoir un effet important sur l'oxydation des hydrocarbures.

Pour prendre en compte les effets physiques de la postoxydation, tels que la turbulence et les effets de mélange, ce schéma cinétique doit être ensuite couplé à un modèle de combustion turbulente adapté à l'utilisation dans un code CFD 3D moteur. Le couplage direct de la cinétique détaillée avec un modèle de combustion CFD n'est pas envisageable à cause du coût en temps de calcul d'une simulation tridimensionnelle d'un écoulement turbulent complexe dans une géométrie complexe avec un grand nombre d'espèces. L'utilisation d'une tabulation de la chimie détaillée permet de réduire considérablement le temps de calcul tout en décrivant la plupart des phénomènes liés à la chimie détaillée. Une technique de tabulation de la chimie sera donc développée et appliquée dans le cadre de ce travail.

Ce manuscrit est structuré de la façon suivante :

- Le chapitre 2 décrit une étude bibliographique de la cinétique d'oxydation des hydrocarbures à basse température et également de l'impact des polluants et des espèces majoritaires présentes dans les gaz brûlés.

- Le chapitre 3 présente les modèles cinétiques existants pour l'oxydation des mélanges PRF/toluène, que nous avons choisis pour représenter l'essence, ainsi que des modèles représentant l'impact des NO_x sur l'oxydation des hydrocarbures.

- Le chapitre 4 décrit le développement d'un schéma cinétique détaillé pour les réactions de postoxydation. Ce schéma, qui décrit l'oxydation d'un carburant ternaire constitué de n-heptane, d'iso-octane et de toluène en présence de NO_x, est validé sur la base d'un nombre important de résultats expérimentaux obtenus sur une large gamme de conditions opératoires.

- Dans le chapitre 5, le modèle cinétique est exploité lors d'études cinétiques réalisées dans les conditions thermochimiques de la postoxydation. L'impact des espèces HCHO, CO, CO_2 et H_2O sur l'oxydation des hydrocarbures est analysé en détail et les limites d'auto-inflammation des gaz brûlés en fonction des conditions thermodynamiques (T, P, dilution et richesse) sont testées.

- Le chapitre 6 explique les bases de la méthodologie de tabulation « Flame Prolongation of ILDM » (FPI) et montre les problèmes posés par l'utilisation d'une tabulation de chimie complexe couplée avec la CFD pour l'application de postoxydation.

- Le chapitre 7 décrit le couplage de la tabulation du schéma cinétique avec le code CFD « IFP-C3D » et le modèle de combustion turbulente ECFM3Z implanté.

- Le chapitre 8 présente des simulations réalisées avec le code CFD en utilisant le modèle de postoxydation développé.

- Enfin, le chapitre 9 contient les conclusions et les perspectives de ce travail.

Les 3 premiers chapitres de ce de ce travail sont écrits en langue française, alors que les autres chapitres sont rédigés en langue anglaise.

Chapitre 2

L'oxydation des hydrocarbures dans un régime de postoxydation

Ce chapitre présente un aperçu de la chimie de combustion des hydrocarbures dans les conditions thermochimiques de la postoxydation. Un résumé bibliographique de la combustion à basse température des alcanes et d'un composé aromatique, le toluène, est présenté. Les étapes-clés de l'oxydation des hydrocarbures à basse température sont rappelées.

L'intérêt est également porté sur les études traitant des interactions des hydrocarbures avec d'autres composants potentiels des gaz brûlés, tels que CO_2, H_2O, CO et les NO_x. Les effets de NO sur l'oxydation des hydrocarbures ainsi que les réactions « clés » de ces interactions seront explicitées. L'objectif de ce travail bibliographique est de mettre en évidence les réactions chimiques importantes.

2.1 Les réactions d'oxydation des hydrocarbures dans les conditions de la postoxydation

Il a été mentionné que la postoxydation se produit dans un régime de température basse et intermédiaire ($T < 1000$ K). La cinétique d'oxydation des hydrocarbures à de telles températures est complexe. Nous décrivons ci-après les réactions habituellement considérées pour modéliser l'oxydation des alcanes et du toluène.

2.1.1. L'oxydation des alcanes

Dans le schéma général d'oxydation des alcanes (Morley et Walker, 1997), on peut distinguer trois zones selon le domaine de température :

1. Le domaine de basse température ($T < 650$ K), dans lequel l'oxydation est gouvernée par l'addition d'oxygène et par les réactions des radicaux péroxyalkyles obtenus (ROO•).

2. Le domaine d'oxydation à haute température ($T > 1000$ K) gouverné par la décomposition des radicaux alkyles par β-scission et par la chimie des radicaux de petite taille.

3. Le domaine de température intermédiaire ($650 < T < 1000$ K) gouverné par l'oxydation des radicaux alkyles pour donner l'alcène conjugué et les radicaux •OOH, grâce à la chimie des radicaux •OOH et de H_2O_2, et à la combinaison des deux mécanismes précédents.

2.1.1.1.   L'oxydation des alcanes à basse température ($T < 650$ K)

L'oxydation dans ce domaine de températures est relativement lente et faiblement exothermique. Cette phase correspond aux régimes d'oxydation lente et de flammes froides qui sont observés dans certaines configurations expérimentales. Le processus d'amorçage prépondérant aux basses températures est l'arrachage d'un atome d'hydrogène de la molécule d'hydrocarbure par une molécule d'oxygène. Un radical alkyle (R•) et un radical hydropéroxyle (•OOH) sont formés :

$$(5) \quad RH + O_2 \leftrightarrow R\bullet + \bullet OOH$$

Le radical R• peut s'additionner sur une molécule O_2 pour donner un radical péroxyalkyle, ROO• :

$$(6) \quad R\bullet + O_2 \leftrightarrow ROO\bullet$$

Puis les radicaux péroxyalkyles s'isomérisent en radicaux hydropéroxyalkyles (•QOOH) par transfert interne d'un atome d'hydrogène en passant par un état de transition cyclique :

$$(7) \quad ROO\bullet \leftrightarrow \bullet QOOH$$

Ce radical •QOOH peut se décomposer en espèces moléculaires (éthers cycliques, alcènes, aldéhydes) et en petits radicaux •X, principalement •OH :

$$(8) \quad \bullet QOOH \rightarrow \bullet X + produits$$

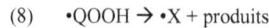

Ce même radical peut aussi s'additionner sur une molécule O_2 pour former un radical hydropéroxypéroxyalkyle qui peut à son tour s'isomériser, puis se décomposer pour former une molécule hydropéroxyde (OQ'OOH) et un radical •OH :

$$(9) \quad \bullet QOOH + O_2 \leftrightarrow \bullet OOQOOH$$

$$(10) \quad \bullet OOQOOH \rightarrow \bullet OH + OQ'OOH$$

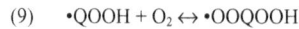

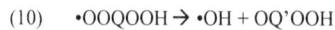

La liaison O-OH dans la molécule hydropéroxyde (OQ'OOH) peut facilement se rompre et conduire à la formation de deux nouveaux radicaux :

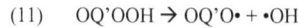

$$(11) \quad OQ'OOH \rightarrow OQ'O\bullet + \bullet OH$$

Cette réaction de ramification (branchement dégénéré) va conduire à la multiplication du nombre de radicaux et permettre d'obtenir une vitesse d'oxydation importante à basse température. Les petits radicaux •X obtenus, en particulier •OH et •OOH, peuvent réagir par métathèse sur le réactif et conduire ainsi à une réaction en chaînes :

$$(12) \quad \bullet X + RH \leftrightarrow XH + R\bullet$$

Un processus de ramification dans une réaction en chaînes peut conduire à une augmentation exponentielle du nombre de radicaux. Cet effet, couplé à l'exothermicité de la réaction, peut conduire à un phénomène d'auto-inflammation. Il est important de noter que ce type de mécanisme, constitué d'une succésion de réactions génériques, ne s'applique qu'à l'oxydation des alcanes constitués d'au moins 3 atomes de carbone.

2.1.1.2. L'oxydation des alcanes à haute température ($T > 1000$ K)

Westbrook et Dryer (1984), comme Warnatz (1984) ont étudié la cinétique d'oxydation des alcanes à haute température. A ces températures, la décomposition par β-scission des radicaux alkyles devient plus rapide que l'addition d'oxygène moléculaire par la réaction (6). Cette décomposition se fait par des réactions du type :

$$(13) \quad R\bullet \to R'\bullet + \text{oléfine}$$

où $R'\bullet$ est un radical alkyle dont le nombre de carbone est inférieur au nombre de carbone de $R\bullet$ ou peut être l'atome $\bullet$H. Par décompositions successives, on obtient principalement des alcènes et des radicaux $\bullet$H, $\bullet CH_3$ et $\bullet C_2H_5$ Les radicaux méthyles sont ensuite essentiellement consommés par des réactions de terminaison ou par réaction avec l'oxygène. Une voie de ramification favorisée à haute température est la dissociation de H_2O_2 en deux radicaux $\bullet$OH.

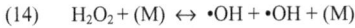

$$(14) \quad H_2O_2 + (M) \leftrightarrow \bullet OH + \bullet OH + (M)$$

M représente tous les partenaires de collision des réactifs. Une autre voie de ramification importante est la consommation des radicaux $\bullet$H par la voie :

$$(15) \quad \bullet H + O_2 \leftrightarrow \bullet OH + \bullet O\bullet$$

qui prend le pas sur la réaction :

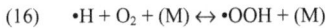

$$(16) \quad \bullet H + O_2 + (M) \leftrightarrow \bullet OOH + (M)$$

La réaction (16) produit des radicaux $\bullet$OOH peu réactifs et gouverne la réactivité en dessous de 1000 K. Grâce aux réactions de ramification (14) et (15), la réactivité globale redevient importante à haute température. Un schéma représentatif de l'oxydation des radicaux alkyles à basse et à haute température est présenté sur la Figure 2-1.

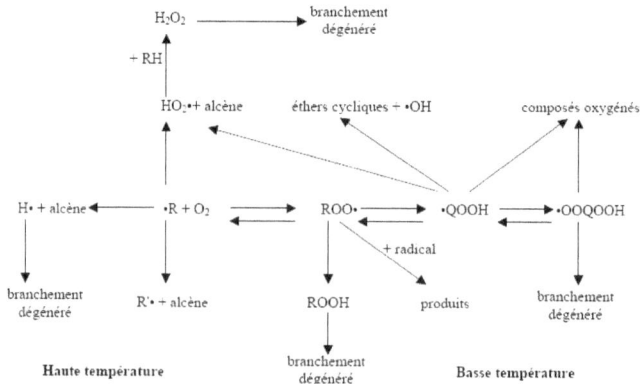

Figure 2-1   Oxydation des radicaux alkyles à haute et à basse température (Touchard, 2005).

2.1.1.3.   L'oxydation des alcanes aux températures intermédiaires

En général on considère que le domaine de basse température s'applique en dessous de 650 K, tandis que la zone de validité de la cinétique de haute température n'est atteinte qu'au-dessus de 1000 K. Entre ces deux domaines de température, les deux mécanismes précédents interviennent et la chimie est gouvernée par les réactions des radicaux •OOH et de H_2O_2. C'est dans cette zone de températures intermédiaires qu'apparaît le phénomène de coefficient négatif de température (NTC) qui est caractérisé par une diminution de la réactivité chimique lorsque la température augmente. L'origine de ce phénomène est liée au fait que la vitesse de la réaction (17) (réaction d'oxydation),

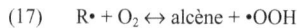

$$(17) \qquad R\bullet + O_2 \leftrightarrow \text{alcène} + \bullet OOH$$

devient importante par rapport à celle de l'addition d'oxygène par la réaction (6) qui est thermodynamiquement défavorisée lorsque la température augmente. A basse température, les radicaux •OOH sont très peu réactifs et réagissent principalement par terminaison :

$$(18) \qquad \bullet OOH + \bullet OOH \leftrightarrow H_2O_2 + O_2$$

La réaction (17) correspond ainsi à une pseudo-terminaison. La compétition entre les réactions (6) et (17) et entre les réactions (8) et (9) entraîne une diminution de la formation des composés hydropéroxydes, du taux de ramification et du taux de formation des radicaux •OH en faveur de la formation des radicaux •OOH et de H_2O_2. Ainsi, la cinétique globale de consommation du réactif est diminuée dans la zone de température intermédiaire ($\approx 700 < T[K] < \approx 1000$) pour les raisons suivantes :

- La vitesse des réactions de ramification (réactions de branchements dégénérés) à basse température diminue et un mécanisme de réactions en chaînes droites (propagation) est établi.

- Le radical •OOH est moins réactif que le radical •OH, et de ce fait la vitesse de propagation est réduite.

- Les réactions faisant intervenir les radicaux •OH sont très exothermiques alors que les réactions avec •OOH sont presque thermoneutres.

- La formation du radical •OOH est favorisé .

Il faut que la température continue à augmenter pour qu'un nouveau processus de ramification par les réactions (14) et (15) se mette alors en place et induise à nouveau une augmentation de la réactivité.

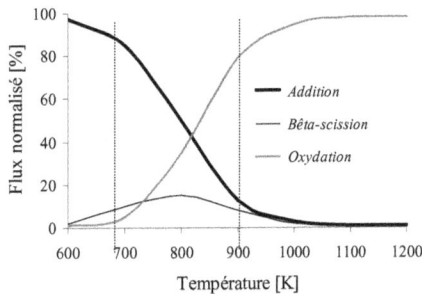

Figure 2-2   Évolution typique avec la température des grands types de réaction mises en jeu lors de l'oxydation des alcanes (d'après Glaude, 1999).

La Figure 2-2 présente les voies principales de réaction des radicaux alkyles, les additions d'oxygène, les oxydations et les béta-scissions, dans les différents régimes de température. Aux températures inférieures à 700 K, les additions d'oxygène dominent largement les oxydations et les béta-scissions. Entre 700 K et 900 K, l'importance des oxydations est maximale. Aux températures supérieures à 900 K, environ 100% des radicaux alkyles sont décomposés par béta-scissions et les additions d'oxygène, ainsi que les oxydations deviennent de moindre importance pour la réactivité globale.

2.1.2. L'oxydation des composés aromatiques – Le toluène

Les espèces aromatiques sont bien connues pour être des composés qui, seuls ou en mélange avec un autre carburant, sont résistantes au cliquetis dans les moteurs à allumage commandé. Le cycle aromatique est plus stable que la chaîne hydrocarbonée des alcanes, ce qui explique que les composés aromatiques brûlent moins facilement. Divers travaux expérimentaux ont été menés sur l'oxydation du toluène. Brezinsky et al. (1984) ont réalisé des expériences en réacteur tubulaire à une température proche de 1200 K. Dagaut et al. (2002) et Bounaceur et al. (2005) ont utilisé un réacteur autoagité pour étudier l'oxydation du toluène. Gauthier et al. (2004), Sivaramakrishnan et al. (2004), Dagaut et al. (2005), Bounaceur et al. (2005), Davidson et al. (2005) et Herzler et al. (2007) ont effectué des expériences en tube à onde de choc. Roubaud et al. (2000), Tanaka et al. (2003b) et Mittal et al. (2007) ont utilisé une machine à compression rapide pour déterminer les délais d'auto-inflammation du toluène. Vanhove et al. (2006) ont également déterminé les délais d'auto-inflammation par une machine à compression rapide pour des mélanges toluène/n-heptane, toluène/iso-octane et toluène 1-hexene. Quant à Dubreuil et al. (2006), ils ont étudié l'oxydation d'un mélange n-heptane/toluène en réacteur auto-agité. Des voies d'oxydations du toluène ont été proposées par plusieurs auteurs, selon lesquels le toluène réagit par métathèse ou arrachage d'un atome d'hydrogène par différents radicaux en produisant principalement des radicaux benzyles et pour une moindre part méthylphényles. En particulier Brezinsky et al. (1984) et Bounaceur et al. (2005) ont écrit les réactions suivantes :

(19)   toluène + R• ↔ benzyl• + RH

(20)   toluène + R'• ↔ •C₆H₄CH₃ + R'H

où R• et R'• représentent des radicaux faisant partie du milieu réactionnel. La formation de benzène ou d'un radical crésoxy par les voies suivantes d'ipso-addition a été   aussi considérée :

(21)   toluène + •H ↔ C₆H₆ + •CH₃

(22)   toluène + •O• ↔ •OC₆H₄CH₃ + •H

Les métathèses produisant les radicaux benzyles qui sont stabilisés par résonance ont un impact inhibiteur sur la cinétique globale. Les métathèses formant les radicaux méthylphényles ont par contre un effet accélérateur. Ces derniers radicaux réagissent surtout par réaction avec l'oxygène par un processus concerté de ramification :

(23)   •C₆H₄CH₃ + O₂ ↔ •OC₆H₄CH₃ + •O•

en produisant des radicaux crésoxy et un atome d'oxygène. Les principaux radicaux formés à partir du toluène sont les radicaux benzyles, méthylphényles et crésoxy. Les radicaux benzyles réagissent principalement par terminaison avec les atomes •O• et •H ou les radicaux X• (•OH, •CH₃ et •OOH), ou avec eux-mêmes.

(24)   benzyl• + •O• ↔ •C₆H₅ + HCHO

(25)   benzyl• + •O• ↔ C₆H₅CHO+ •H

(26)   benzyl• + •H ↔ toluène

(27)   benzyl• + X• ↔ $C_6H_5CH_2X$

(28)   benzyl• + •benzyl ↔ bibenzyl

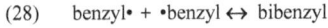

Ces réactions diminuent la réactivité globale du système. Aux températures en dessous de 1000 K, les radicaux crésoxy, stabilisés par résonance, réagissent surtout avec des radicaux •OOH (Dagaut et al. 2002) ou par arrachage d'un atome d'hydrogène sur le phénol en produisant des crésols (Andrae et al. 2007).

(29)   •$OC_6H_4CH_3$ + •OOH ↔ $HOC_6H_4CH_3$ + O_2

(30)   •$OC_6H_4CH_3$ + C_6H_5OH ↔ $HOC_6H_4CH_3$ + C_6H_5O•

Bounaceur et al. (2005) ont noté qu'à basse température, la réaction (29) a un impact très inhibiteur sur la cinétique globale. La Figure 2-3 montre le schéma d'oxydation du toluène comme il a été déduit par ces auteurs d'une analyse de vitesse à une température de 893 K.

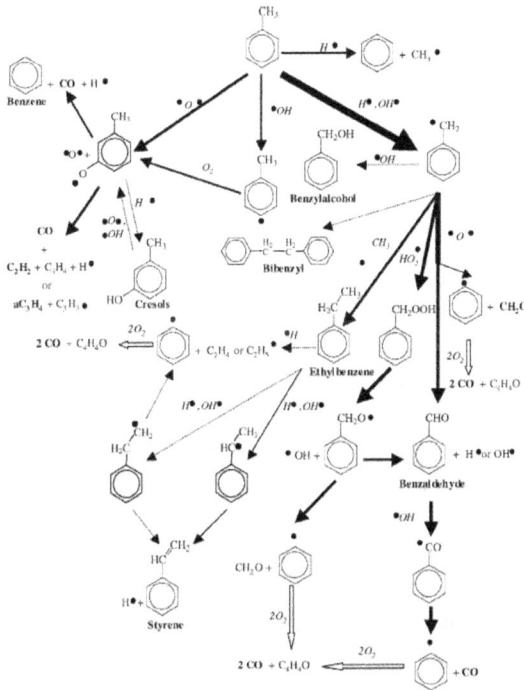

Figure 2-3   Schéma d'oxydation du toluène à une richesse φ = 0.5, pression atmosphérique et une température de 893 K après un temps de séjour de 4 s dans un réacteur auto-agité (Bounaceur et al. 2005).

2.2 L'impact des composants des gaz brûlés sur l'oxydation des hydrocarbures

CO_2 et H_2O, qui sont les produits finaux d'une combustion parfaite d'un hydrocarbure avec l'oxygène, sont généralement considérés comme des espèces inertes qui ne participent à aucune réaction chimique. On leur attribue avant tout un effet diluant sur la cinétique d'oxydation des hydrocarbures. CO est le produit d'une combustion incomplète déficitaire en oxygène (richesse $\phi > 1$) ou de la dissociation de CO_2 aux hautes températures.

Présent à 79 % (pourcentage molaire) dans l'air, l'azote constitue naturellement le diluant de référence en combustion. L'azote, en tant que molécule diatomique, se distingue surtout par ses caractéristiques thermiques par rapport à des molécules triatomiques comme CO_2 et H_2O. La Figure 2-4 compare la capacité calorifique (C_p) des molécules diatomiques, N_2 et CO, à celles des molécules triatomiques, CO_2 et H_2O, en fonction de la température.

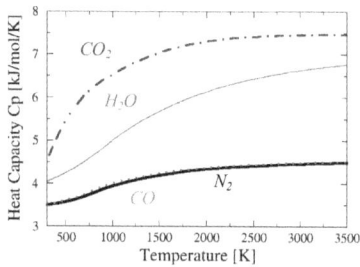

Figure 2-4   Capacité calorifique (C_p) des espèces N_2, CO, CO_2 et H_2O (NIST).

La Figure 2-4 montre que les C_p des molécules triatomiques CO_2 et H_2O, sont supérieurs à ceux des molécules diatomiques, N_2 et CO, qui dans leur cas sont quasiment identiques. La présence de CO_2 et de H_2O dans les gaz brûlés impose donc une augmentation du C_p global par rapport à un mélange des mêmes réactifs qui serait dilué uniquement par N_2. Une cinétique ralentie pourrait en être la conséquence. Plusieurs travaux comparent les effets en tant que diluants de CO_2 et H_2O avec celui de N_2 et certains montrent un effet cinétique de ces composés.

2.2.1. L'impact du CO

La plupart des travaux étudiant l'impact de CO sur l'oxydation de combustibles concerne l'oxydation des mélanges CO/H_2. Les premières études sur le système de réactifs $CO/H_2/O_2$ ont été effectuées par Dixon et al. (1977), Gardiner et Olson (1980), et Westbrook et Dryer (1984). Dans les années 90, Yetter et al. (1991) ont proposé un mécanisme d'oxydation de mélanges CO/H_2 en réacteur à écoulement piston. Ce mécanisme a ensuite été mis à jour et sur cette base, un mécanisme d'oxydation du méthanol et du méthane a été développé (Mueller et al. 1999 ; Li et al. 2004). Le groupe de Li a également publié un mécanisme pour la combustion de CO, HCHO et CH_3OH dans des conditions d'oxydation : flammes laminaires, tubes à onde de choc et réacteur à écoulement piston (Li et al. 2007). Davis et al. (2005) ont développé un mécanisme de l'oxydation de mélanges CO/H_2 en optimisant les paramètres cinétiques des réactions élémentaires ainsi que les données thermodynamiques et de transport. Frassoldati et al. (2007) et Cuoci et al. (2007) ont modifié un mécanisme développé par Ranzi et al. (1994) pour reproduire l'oxydation des mélanges CO/H_2. Leur mécanisme a été testé dans une large gamme de conditions expérimentales. Le sous-mécanisme d'oxydation de mélanges CO/H_2 du mécanisme d'oxydation du méthane de Leeds (Hughes et al. 2001) a été analysé par Zsély et al. (2005) qui l'ont mis à jour et testé pour l'oxydation d'un mélange H_2 et CO en présence d'eau. Le Cong a étudié en 2007 l'impact de CO sur l'oxydation de H_2 dans un réacteur autoagité en présence d'eau (Le Cong, 2007).

Très peu d'études ont été effectuées sur l'impact du CO sur l'oxydation des hydrocarbures. Moréac (2003) a effectué des expériences sur l'impact de CO sur l'oxydation de méthanol, de n-heptane, d'iso-octane et de toluène dans un réacteur auto-agité sans constater d'effet considérable. Subramanian et al. (2007) ont effectué une étude numérique et ont analysé l'impact de CO sur l'auto-inflammation de n-heptane. Ils ont testé différents mécanismes détaillés et constaté un impact complexe de CO, sur l'oxydation de cet alcane, en fonction de la température. A une température de 600 K, l'addition de 2 mol% augmente le délai d'auto-inflammation jusqu'à 8 %, alors qu'à une température de 1000 K, les délais d'auto-inflammation sont réduits d'environ 20 %. Machrafi et al. (2008) ont effectué des expériences dans un moteur HCCI en testant l'impact de CO sur les délais d'auto-inflammation d'un carburant PRF (Primary Reference Fuel). Ils ont ajouté des concentrations de CO jusqu'à 170 ppm sans détecter d'effets notables. De plus, ils ont étudié numériquement l'impact de l'ajout de concentrations de CO plus élevées et observé un impact inhibiteur pour l'addition de 1 mol% de CO.

2.2.2. L'impact du CO_2

Lai et Thomas (1995) ont analysé l'effet de CO_2 sur la combustion de H_2, de l'éthane (C_2H_6) et du méthane (CH_4) dans l'air dans un réacteur fermé ($T = 800$-1300 K). En remplaçant N_2 par CO_2, ils n'ont pas constaté d'effet cinétique.

Le Cong (2007) a étudié expérimentalement l'impact de CO_2 sur l'oxydation de mélanges CH_4/H_2, dans un réacteur auto-agité. Il a analysé la cinétique à des températures variant entre 925 et 1450 K et constaté un impact inhibiteur de la présence de CO_2. Saso (2002) a étudié numériquement le potentiel de différents inhibiteurs chimiques lors de la combustion du méthane en utilisant le mécanisme GRI-MECH2.11 (Smith 1994). Il a calculé des vitesses de flamme laminaire en présence de différents inhibiteurs. Il a conclu que CO_2 interagit principalement par ses propriétés thermiques et il lui a donc attribué un rôle inhibiteur.

De même, Liu et al. (1983) ont étudié numériquement l'impact de la substitution de N_2 par CO_2 dans une flamme de diffusion d'éthylène (C_2H_4). Ils ont différencié l'addition de CO_2 dans l'oxydant et dans le combustible et ils ont pu constater un impact chimique de CO_2 dans les deux cas qui se superpose à son effet thermique. Lorsque le CO_2 est injecté avec le carburant, l'impact thermique dû à son C_p élevé surpasse légèrement son impact cinétique. Dans le cas de l'injection de CO_2 du côté de l'oxydant, son impact cinétique domine l'impact thermique. Glarborg et Bendtsen (2008) ont effectué des expériences sur l'oxydation du méthane dans un réacteur tubulaire à forte dilution pour des températures de 1200 et de 1800 K en augmentant la substitution de N_2 par CO_2 de 0.76 à 95%. Ils ont constaté une forte augmentation de la production de CO en présence de CO_2.

2.2.3. L'impact de l'eau - H_2O

La plupart des études consacrées aux effets de l'addition d'H_2O sur la combustion d'un combustible a été effectuée pour des flammes d'hydrogène moléculaire. Lai et Thomas (1995) ont analysé l'effet de H_2O sur la combustion de H_2, C_2H_6 et CH_4. En substituant N_2 par H_2O, ils n'ont pas pu constater d'impact cinétique notable. Liu et Mac Farlane (1983) ont comparé les vitesses de flamme laminaire pour des flammes prémélangées H_2/air et H_2/air/H_2O. En variant les concentrations de H_2O de 0 à 15 %, ils ont observé des vitesses de flamme ralenties en augmentant la teneur en H_2O. Koroll et Mulpuru (1986) ont confirmé les observations de Liu et Mac Farlane (1983) en constatant des vitesses de flamme réduites résultant d'un taux de dilution augmenté. Néanmoins, ils ont observé des vitesses de flammes accélérées lors d'une substitution de N_2 par H_2O.

Lamoureux et al. (2002) ont mesuré les vitesses de flamme de mélanges H_2-air dilués avec différentes quantités d'eau (10%, 20% et 30%) en milieu pauvre. Ces résultats ont permis de conclure à un effet inhibiteur de l'eau sur les vitesses de flamme d'hydrogène. Pelett et al. (1992) ont étudié les conditions d'extinction d'une flamme de diffusion H_2-air à contre courant. Ils ont observé des limites d'extinction plus élevées lors d'une substitution de N_2 par H_2O sur le côté oxydant.

Par l'analyse cinétique dans un statoréacteur à combustion supersonique (scramjet combustion), Mitani et al. (1995, 1997) ont vérifié que l'eau inhibe l'auto-inflammation de H_2 et influe fortement sur les réactions de troisième corps. Wang et al. (2003) ont déterminé des délais d'auto-inflammation de mélanges contenant de l'hydrogène. Ils ont effectué plus de 150 expériences à des températures variant entre 900 et 1350 K. Ils ont conclu que le délai d'auto-inflammation dépend fortement de la quantité d'eau ajoutée et de la température. Le Cong (2007) a étudié expérimentalement l'impact de H_2O sur l'oxydation du méthane, dans un réacteur auto-agité. Il a analysé la cinétique à des températures variant entre 1025 et 1450 K et a constaté un impact inhibiteur de la présence de H_2O.

Les études de la littérature analysant l'impact de CO_2 et H_2O sur l'oxydation de combustibles portent surtout sur des flammes de H_2 et concernent des températures supérieures à 900 K. On ne trouve guère d'études analysant l'impact de CO_2 et de H_2O dans les conditions thermochimiques de la postoxydation (basses températures et hydrocarbures contenant plusieurs atomes de carbone).

2.2.4. L'impact des NO_x

Les oxydes d'azote sont des composés qui interagissent de manière notable sur l'oxydation des hydrocarbures, même en faibles quantités. Divers auteurs ont pu observer des effets accélérateurs ou inhibiteurs sur l'oxydation de différents combustibles (Dubreuil et al. 2006,   Glaude et al. 2005,   Moréac et al. 2003,   Frassoldati et al. 2003, Alzueta et al. 2001, Alzueta et al. 1997). L'ampleur et la nature de ces effets varient en fonction de la température et de la pression, ainsi qu'en fonction de la teneur en NO (Moréac et al. 2003).

Plusieurs études ont été effectuées sur l'oxydation des hydrocarbures en présence de NO, à basse température (Moréac et al. 2003,   Glaude et al. 2005 Dubreuil et al. 2006,). Ces différentes études montrent, que de faibles quantités de NO (quelques dizaines de ppm) peuvent avoir un effet accélérateur sur l'oxydation d'hydrocarbures, tels que le n-heptane par exemple. Cependant, au-delà d'une certaine quantité, l'addition croissante de NO aboutit à un effet inhibiteur (Moréac, 2003). Diverses réactions ont été proposées pour expliquer ces observations. Pour les effets accélérateurs de NO, la littérature met en évidence le rôle des réactions suivantes (Moréac et al. 2003) :

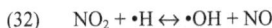

(31) $NO + \bullet OOH \leftrightarrow NO_2 + \bullet OH$

(32) $NO_2 + \bullet H \leftrightarrow \bullet OH + NO$

Aux températures intermédiaires, la production du radical •OOH devient très importante. L'effet promoteur de NO dans ces conditions est ainsi expliqué par la réaction (31). Les radicaux •OOH sont alors transformés en radicaux •OH qui sont beaucoup plus réactifs avec les hydrocarbures et donc accélèrent leur oxydation. La réaction (32) est aussi proposée par certains auteurs, car elle permet la régénération de NO pendant l'oxydation des hydrocarbures. L'influence cinétique des deux réactions consécutives (31) et (32) est que l'on transforme respectivement l'atome •H et le radical •OOH, peu réactif, en un radical, •OH, plus réactifs. Cet effet catalyseur de NO est indépendant de l'hydrocarbure initial, mais son intensité est variable. La réaction (32) devient moins importante en excès d'oxygène du fait de la consommation croissante des atomes H• par la réaction :

$$(16) \qquad H\bullet + O_2 + M \leftrightarrow \bullet OOH + M$$

L'effet de NO à basse température pourrait donc être moins important en excès d'oxygène. Pour expliquer les effets inhibiteurs de NO à basse température, quatre réactions ont été proposées : (Glaude et al. 2005; Alzueta et al. 2001, Alzueta et al. 1997; Prabhu et al. 1996) :

$$(33) \qquad ROO\bullet + NO \leftrightarrow RO\bullet + NO_2$$

$$(34) \qquad NO + \bullet OH + M \leftrightarrow HONO + M$$

$$(35) \qquad R\bullet + NO_2 \leftrightarrow RNO_2$$

$$(36) \qquad RO\bullet + NO_2 \leftrightarrow RONO_2$$

En présence d'une grande quantité de NO (de l'ordre de 400 ppm), un effet inhibiteur sur l'oxydation de différents hydrocarbures à des températures inférieures à 700 K a été observé. La réaction (33) transforme des radicaux ROO• en radicaux RO• et est en concurrence avec l'addition d'oxygène qui conduit aux réactions de ramification (cf. section 2.1.1). Il y a donc un impact inhibiteur de la réaction (33) sur l'oxydation des alcanes à basse température. Cette inhibition est aussi expliquée par la formation d'acide nitreux (HONO) via la réaction (34). L'acide nitreux est une espèce stable dont la formation consomme des radicaux •OH. Une autre réaction inhibitrice citée est la recombinaison de NO_2 avec un radical alkyle, réaction (35), pour former un nitroalcane. Enfin, une dernière voie d'inhibition peut provenir de la recombinaison de NO_2 avec des radicaux oxygénés par des réactions de terminaison, réaction (36).

2.3   Conclusions sur la recherche bibliographique

La bibliographique sur la cinétique d'oxydation des hydrocarbures dans les conditions thermochimiques de la postoxydation a montré que la cinétique des réactions mises en jeu est complexe. Le régime de postoxydation couvre la plage de température où peut être observée la zone de « *Coefficient Négatif de Température* » (NTC). Les hydrocarbures imbrûlés sont principalement composés d'espèces aliphatiques et aromatiques, deux familles d'hydrocarbures qui s'oxydent de manières différentes. Lors de l'oxydation des alcanes, on observe une importante réactivité dès 650 K et un effet de NTC pour des températures comprises entre 700 et 850 K. Ce comportement a été largement étudié dans la littérature. En ce qui concerne les espèces aromatiques faiblement substituées, aucune zone de NTC marquée n'a pu être observée. Ces espèces sont en outre beaucoup plus résistantes à l'oxydation et s'auto-enflamment à des températures plus élevées (~900 K) que les alcanes. Par conséquent, leur oxydation n'a pas pu être étudiée à aussi basse température que dans le cas des espèces aliphatiques et les interactions à basses températures entre espèces aliphatiques et aromatiques restent mal comprises.

Les gaz brûlés sont caractérisés par une forte présence d'espèces habituellement considérées comme des diluants, tels que N_2, H_2O et CO_2. La recherche bibliographique révèle que l'impact de ces espèces sur l'oxydation des hydrocarbures ne se limite pas toujours à uniquement un effet de dilution, mais qu'un effet cinétique de H_2O et CO_2, entre autres, a pu être observé également. Peu d'études ont été effectuées concernant l'impact du CO, même si un possible impact de ce composé sur la cinétique d'oxydation des hydrocarbures a été suggéré. Beaucoup plus d'études ont été consacrées à l'effet de la présence de NO lors de l'oxydation des hydrocarbures. Un impact cinétique complexe de NO été constaté, même quand il est présent en faibles concentrations (~50 ppm). Néanmoins la majorité des travaux étudiant l'impact de NO sur l'oxydation des hydrocarbures s'est limitée au cas des alcanes. L'impact de NO sur l'oxydation des espèces aromatiques reste, ainsi, très peu connu.

Chapitre 3

Les modèles d'oxydation du n-heptane, de l'iso-octane et du toluène, de leurs mélanges et de leurs interactions avec les NO$_x$

Le travail de modélisation cinétique décrit dans ce mémoire repose sur l'utilisation de mécanismes d'oxydation de carburant modèle. Un carburant modèle est constitué d'hydrocarbures simples qui dans leur ensemble ont des caractéristiques physicochimiques proches d'un carburant réel. Le choix du carburant modèle et des espèces « clés » pour reproduire son oxydation est donc primordiale pour modéliser la cinétique chimique dans les conditions de postoxydation.

Ce chapitre discute d'abord notre choix du carburant modèle, (composé de n-heptane, d'iso-octane et de toluène) et des espèces« clés ». Ensuite nous presentons les principaux mécanismes détaillés disponibles dans la littérature pour l'oxydation du n-heptane, de l'iso-octane et du toluène et de leurs mélanges. Les mécanismes décrivant l'oxydation de ces composés comme réactif organique unique seront décrits en premier. Ces modèles d'oxydation sont à la base des modèles d'oxydation des mélanges binaires et ternaires d'hydrocarbures. En général ceux-ci sont obtenus par un couplage entre les modèles d'oxydation de différents hydrocarbures. Les principaux mécanismes d'oxydation des mélanges binaires contenant du n-heptane, de l'iso-octane ou du toluène seront présentés avant de porter notre intérêt sur les mécanismes descriptifs des mélanges ternaires du type PRF/toluène. Ce chapitre se termine par une présentation de deux modèles d'oxydation décrivant les interactions entre les NO_x et les hydrocarbures.

3.1 Espèces « clés » nécessaires pour une modélisation cinétique de la postoxydation

Les carburants automobiles sont en général constitués de centaines d'hydrocarbures principalement répartis entre les familles des alcanes, des oléfines et des composés aromatiques. Le tableau 1 présente la composition moyenne d'une essence européenne commerciale (Guibet, 1997).

Tableau 1   Composition d'une essence européenne commercial (Guibet, 1997).

Nombre d'atomes de carbone	Composés saturés				Composés insaturés		Total
	Alcanes linéaires	Alcanes ramifiés	Ethers	Alcanes cycliques	Alcènes	Composés aromatiques	
	[% mass.]	[% mass.]	[% mass.]	[% mass.]	[% mass.]	[% mass.]	[% mass.]
4	5,14	0,30			1,49		6,93
5	1,26	7,84	0,50		10,11	0,50	19,71
6	0,64	6,34	3,00	1,19	5,07	1,23	17,47
7	0,65	3,22		1,05	1,56	8,11	14,59
8	0,48	11,47		0,43	0,34	13,61	26,33
9	0,11	1,12		0,16	0,07	9,49	10,95
10	0,01	0,09		0,09	0,02	2,80	3,01
11		0,10				0,25	0,35
12		0,61					0,61
13		0,01					0,01
Total	8,29	31,10	3,50	2,92	18,66	35,49	99,96

Aujourd'hui, les biocarburants (éthanol, biodiesel, …) sont aussi de plus en plus intégrés dans les formulations de carburants commerciaux, tant pour l'essence que pour le carburant diesel. Lorsqu'il s'agit de modéliser la chimie, il n'est cependant pas possible de prendre en compte toutes les espèces présentes dans les carburants réels. Des choix de modélisation s'avèrent donc nécessaires. La notion d'espèces « clé » pour la définition d'un modèle cinétique repose ainsi sur l'hypothèse que la modélisation de la réactivité d'un mélange composé de telles espèces permettra de reproduire le comportement cinétique du carburant dit réel. Dans le cas de la postoxydation, les espèces « clés » doivent permettre de reproduire le comportement cinétique des gaz brûlés. Les composants des gaz brûlés peuvent être classés en 4 groupes :

- *Le carburant initial et l'oxygène,*
- *Les produits de combustion CO$_2$, H$_2$O et N$_2$,*
- *Les espèces partiellement oxydées,*
- *Les NO$_x$.*

Les groupes de composants cités ci-dessus seront décrits de suite.

- *Le carburant initial et l' oxygène :*

La composition d'un carburant modèle doit représenter les principales familles d'hydrocarbures présentes dans le carburant réel (voir tableau 1) car à chaque famille sont associées des propriétés chimiques différentes en fonction de la structure des molécules. Des travaux réalisés récemment à l'IFP pour la modélisation de la cinétique chimique appliquée aux moteurs à essence à allumage commandé (Jay et al. 2006) ont été basés sur l'utilisation d'un mélange ternaire composé de :

- o *N-heptane* qui est à la fois une molécule de référence pour la simulation de la réactivité de carburants diesel dû à son indice de cétane[3] de 56 proche de celui d'un gazole commercial et pour celle des carburants essence car il s'agit d'une des deux molécules de référence pour la mesure des indices d'octane[4] avec une valeur de 0.

- o *Iso-octane* (2,2,4-triméthylpentane) qui est l'autre espèce « clé » pour la caractérisation de l'essence dû à son indice d'octane élevé (100) et qui est la seconde molécule de référence pour la mesure des indices d'octane.

---

[3] *L'indice de cétane évalue la capacité d'un carburant à s'auto-inflammer sur une échelle de 0 à 100. Les molécules de référence caractérisant l'indice de cétane sont l'alpha-méthylnaphtalène (indice 0) et le n-cétane (indice 100). On peut utiliser aussi comme molécule de référence à faible indice de cétane l'heptaméthylnonane (indice 15) qui est un isomère du n-cétane fortement ramifié. L'indice de cétane est particulièrement important pour les <u>moteurs diesel</u> où le carburant doit s'auto-inflammer sous l'effet de la pression et de la température (Guibet, 1997).*

[4] *L'indice d'octane 'RON' (Research Octane Number) caractérise la résistance du carburant à l'auto-inflammation pour les <u>moteurs à essence</u>. Par définition, l'indice d'octane varie entre 0 et 100. Sa valeur est égale à la fraction volumique d'iso-octane contenue dans un mélange iso-octane/n-heptane ayant, dans les conditions du moteur, les mêmes caractéristiques d'auto-inflammation que le composé ou le mélange étudié (Guibet, 1997).*

Le n-heptane et l'iso-octane, les deux molécules de référence pour la mesure des indices d'octane, sont appelés PRF (Primary Reference Fuel).

o *Toluène* représentant la famille des composés aromatiques qui possèdent un rapport C/H élevé (toluène : 0.875) et un fort indice d'octane (toluène : RON = 121) et qui sont présents en quantités importantes dans les essences (voir tableau 1).

Un nombre croissant d'études dans la littérature concerne la modélisation de la combustion de mélanges ternaires composés de PRF (n-heptane/iso-octane) et de toluène (Andrae et al. 2007 ; Pires da Cruz et al. 2007 ; Yahyaoui et al. 2007). Les proportions de chacun des composants d'un mélange représentatif d'une essence peuvent être choisies par exemple en fonction de cibles d'indice d'octane et de rapport C/H. La présence de biocarburants dans le carburant modèle ne sera pas retenue dans le cadre de notre travail même si, en général, la cinétique de la combustion de molécules en espèces C_1-C_2, comme le méthanol et l'éthanol, peut être simulée à l'aide de la plupart des modèles d'oxydation des hydrocarbures. L'oxygène est présent dans les gaz brûlés uniquement dans le cas d'une combustion pauvre, voir stœchiométrique. Lors d'une combustion riche, l'oxygène est totalement consommé.

• *Les produits de la combustion CO_2, H_2O et N_2 :*

Les principaux produits de la combustion, CO_2, H_2O et N_2 sont de leur côté naturellement présents dans les mécanismes cinétiques détaillés. Les principales réactions d'interaction entre ces espèces, le carburant et les intermédiaires réactionnels sont généralement prises en compte dans les schémas cinétiques.

• *Les espèces partiellement oxydées :*

Concernant les interactions du carburant avec les espèces partiellement oxydées, telles que le CO, les alcools, les cétones ou les aldéhydes (< C_3), elles sont, en général, également présentes dans les modèles d'oxydation.

• *Les NO$_x$:*

Les interactions entre les NO$_x$ et les hydrocarbures tels que le n-heptane, l'iso-octane et le toluène ne sont que très rarement intégrées dans les modèles d'oxydation. Leurs réactions restent encore en grande partie à définir.

Dans ce travail, nous avons considéré comme espèces « *clé* » pour une modélisation cinétique de la postoxydation : le *n-heptane,* l'*iso-octane,* le *toluène* et le *monoxyde d'azote.* **Le carburant modèle** choisi pour les travaux numériques présentés dans les chapitres 5.1, 7, et 8 sera **composé de 13,7 mol% de n-heptane, 42,9 mol% d'iso-octane et 43,4 mol% de toluène.** Un tel mélange est caractérisé par un **RON de 95 et un rapport de H/C de 1.8,** proches des valeurs mesurées pour le carburant standard utilisé aux bancs moteurs de l'IFP.

3.1.1. Les modèles d'oxydation de composés purs

Une sélection de schémas cinétiques modélisant l'oxydation du n-heptane, de l'iso-octane et du toluène sera présentée ci-après. Les mécanismes détaillés ont été développés grâce à des résultats expérimentaux qui ont été obtenus en réacteurs à écoulement (FLR), en réacteurs auto-agités (JSR), en tubes à onde de choc (ST) et en machines à compression rapide (RCM). L'équipe de Dagaut (eg. Dagaut et al. 1994) a réalisé de nombreuses études expérimentales en réacteur auto-agité par jets gazeux sur une large gamme de conditions expérimentales ($T = 550$-1150 K, $P = 10$ atm, $0,3 \leq \phi \leq 1,5$). L'équipe de Minetti a réalisé des expériences en machine à compression rapide en observant des délais d'auto-inflammation (par exemple Minetti et al. (1996)). Ces expériences donnent des informations sur la réactivité de chacun des hydrocarbures et identifient les produits intermédiaires majoritaires formés pendant l'oxydation.

3.1.2. Les modèles d'oxydation du n-heptane

Les composés oxygénés intermédiaires stables mesurés au cours de l'oxydation du n-heptane à basse température sont essentiellement des hétérocycles et des aldéhydes : formaldéhyde, acétaldéhyde, propionaldéhyde et butyraldéhyde. Les espèces non-oxygénées sont principalement des alcènes : l'éthylène, le 1-propène, le 1-butène, le 1-pentène, le 1-hexène et les oléfines en C$_7$ (Guibet, 1997). Le Tableau 2 présente différents modèles développés pour l'oxydation du n-heptane.

Tableau 2   Sélection de quelques modèles cinétiques écrits pour l'oxydation du n-heptane pur.

Nombre d'espèces	Nombre de réactions	Réacteur	Validations T [K]	P [atm]	ϕ [−]	Référence
212	765	ST	1000-1500	1	1	Westbrook et al. (1988)
72	519	JSR ST	950-1200 1100-1760	1 3-15	0,2-2	Chakir et al. (1991)
620	2400	ST	600-900	3-40		Chevalier et al. (1992)
109	659	JSR	900-1200	1	0,5-2	Lindstedt et Maurice (1995)
100	2000	JSR ST	550-1000 650-1300	10-40 6-42	0,3-2	Ranzi et al. (1995)
200	1200	ST	650-1300	3-40	0,5-2	Nehse et al. (1996)
42	266	FLR JSR	940-1075 900-1200	3 1	0,8-2.3 0,5-2	Held et al (1997)
544	2446	FLR JSR RCM ST	550-850 550-1000 650-1000 650-1300	12.5 10-40 6-11 6-42	1 0,5-1.5 1 0,5-2	Curran et al. (1998)
273	1282	JSR	550-1000	1-10	0,2-1	Glaude et al. (2000)
360	1817	RCM ST JSR	650-900 650-1250 600-1000	6-11 6-42 10	1 1 1	Buda et al. (2005)

Dans les années 1990, de nombreux auteurs ont proposé des mécanismes d'oxydation du n-heptane. Westbrook et al. (1988), Chakir et al. (1991), Chevalier et al. (1992), Lindstedt et Maurice (1995) et Ranzi et al. (1995) ont ainsi développé de tels schémas cinétiques. Ces modèles ont été validés à partir de données obtenues dans des réacteurs auto-agités et en tubes à onde de choc. Lindstedt et Maurice, (1995) ont également validé leur modèle en flamme de diffusion. Held et al. (1997) ont développé un modèle semi-détaillé d'oxydation de n-heptane qui a largement été utilisé dans la littérature (Davis et al. 1998; Ingemarsson et al. 1999).

Curran et al. (1998) ont proposé un modèle d'oxydation du n-heptane qui a été validé à basse et à haute température. Ce modèle possède l'avantage par rapport aux modèles précédents de présenter une construction systématique selon des catégories de réactions élémentaires et une discussion détaillée du choix des constantes de vitesse. Celles-ci sont établies pour chacune des 25 catégories de processus, sans ajustement des paramètres d'une réaction particulière à l'intérieur des classes. L'approche rigoureuse et systématique de Curran et al. (1998) est proche de la démarche de construction de mécanismes par génération automatique.

La génération automatique garantit une bonne lisibilité des mécanismes en assurant une logique et une cohérence entre les réactions. Glaude et al. (2000) ont proposé un mécanisme d'oxydation du n-heptane généré automatiquement par le logiciel EXGAS (Warth et al. 1998) et qui a été validé sur des résultats obtenus en réacteur auto-agité. Ce mécanisme a par la suite été amélioré par Buda et al. (2005) qui l'a validé sur des résultats obtenus en tube à onde de choc et en machines à compression rapide. Les réactions des produits d'oxydation du n-heptane, tels que les alcènes ou les éthers cycliques, ne sont plus des processus élémentaires, mais des étapes globalisées de décomposition, pour des raisons de taille du modèle.

3.1.3. Les modèles d'oxydation de l'iso-octane

Étant donné que l'iso-octane est une espèce ramifiée, elle possède davantage de liaisons C-H primaires que le n-heptane. L'énergie des liaisons C-H primaires ($E_{C-H\,1} > 60\ \text{kJ·mol}^{-1}$) est plus importante que celle des liaisons C-H secondaires ($E_{C-H\,2} > 50\ \text{kJmol}^{-1}$) ou même tertiaires ($E_{C-H3} > 40\ \text{kJ·mol}^{-1}$). L'oxydation de l'iso-octane est donc moins rapide que celle du n-heptane. De plus, la réactivité dépend aussi de l'étape de ramification. La structure de l'iso-octane défavorise l'isomérisation des radicaux $RO_2^{\bullet}$ et $O_2QOOH^{\bullet}$. Il y a donc moins d'intermédiaires oxygénés produits pendant l'oxydation de l'iso-octane que pendant celle du n-heptane à basse température (Dagaut et al. 1994) ce qui contribue aussi au ralentissement de la réactivité de ce composé branché. Plusieurs équipes ont développé des mécanismes d'oxydation de l'iso-octane dont une sélection est présentée par le Tableau 3.

Tableau 3   Sélection de quelques mécanismes détaillés écrits pour l'oxydation de l'iso-octane.

Nombre d'espèces	Nombre de réactions	Réacteur	Validations			Référence
			T [K]	P [atm]	ϕ [–]	
324	1303	JSR	513-701	1	1	Westbrook et al. (1988)
		ST	1000-1500			
145	2500	FLR	800-1100	1	0,3-1.5	Ranzi et al. (1997)
		JSR	550-1100	10	1,5	
		RCM	650-1300	13-16	1	
		ST	650-1300	36	0,5-2	
273	1282	JSR	550-1000	1-10	0,2-1	Glaude (1999)
860	3600	FLR	600-945	1-12.5	0,05-1	Curran et al. (2002)
		JSR	700-1150	10	0,5-1,5	
		ST	1000-1800	2.1-40	1	
351	1684	RCM	650-900	6-11	1	Buda (2005)
		ST	650-1250	13-45	1	
		JSR	600-1000	10	1	

Un des premiers mécanismes détaillés d'oxydation de l'iso-octane a été développé par Westbrook et al. (1988) pour l'oxydation de l'iso-octane à haute température. Ranzi et al. (1997) et Glaude et al. (1999) ont proposé des mécanismes validés aux basses températures. Quant à Curran et al. (2002), ils ont développé un modèle d'oxydation de l'iso-octane basé sur celui proposé par Westbrook al. (1988). Ce modèle a été construit avec une architecture similaire à celle du modèle proposé pour l'oxydation du n-heptane (Curran et al. 1998) mentionné dans la section précédente et couplé ensuite à celui-ci. Il s'agit donc d'un modèle d'oxydation de mélanges PRF, mais dans la littérature, ce modèle est généralement référencé comme modèle d'oxydation de l'iso-octane et sera donc cité également parmi les modèles d'oxydation de l'iso-octane pur.

La mise à jour du logiciel EXGAS par Glaude et al. (2000) a également permis la génération d'un mécanisme détaillé d'oxydation de l'iso-octane. Glaude et al. (2000) l'ont validé à partir des résultats expérimentaux obtenus en réacteur auto-agité. Buda et al. (2005) ont amélioré ce mécanisme et l'ont validé afin de mieux reproduire des expériences en machines à compression rapide et en tubes à onde de choc. Parmi les mécanismes cités ci-dessus, seuls les modèles de Ranzi et al. (1997), de Glaude et al. (2000) et de Buda et al. (2005) ont été générés automatiquement.

3.1.4. Les modèles d'oxydation du toluène

Un des premiers mécanismes détaillés développés pour l'oxydation du toluène a été proposé par Emdee et al. (1992). Linsdstedt et Maurice (1996) ont également étudié l'oxydation du toluène grâce à un schéma cinétique détaillé. Ces deux modèles ont été validés pour des températures autour de 1200 K en utilisant les résultats expérimentaux obtenus par

Brezinsky et al. (1984) en réacteur tubulaire. Klotz et al. (1998), et plus tard, Sivaramakrishnan et al. (2004), ont amélioré le modèle de Emdee et al. (1992).

Le modèle de Djurisic et al. (2001) reproduit l'oxydation du toluène à des températures supérieures à 1000 K. Leur modèle est validé à partir de résultats obtenus en réacteur tubulaire, en tube à onde de choc et également en flamme prémélangée. Pitz et al. (2001) ont mis au point un modèle d'oxydation du toluène en se basant sur celui proposé par Zhong (1998). Ils l'ont couplé avec le mécanisme d'oxydation du n-heptane de Curran et al. (1998). Ce modèle est validé à partir de résultats experimentaux obtenus en tube à onde de choc et en réacteur tubulaire. Dagaut et al. (2002) ont proposé un mécanisme contenant 120 espèces et 920 réactions qui a été validé grâce à des résultats obtenus en réacteur auto-agité, en tube à onde de choc et en flamme prémélangée (vitesse de flamme). Le mécanisme de Bounaceur et al. (2005) a été validé en utilisant des résultats obtenus en réacteur auto-agité (873 K et 923 K) et en tube à onde de choc (températures entre 1305 K et 1795 K). Les mécanismes détaillés d'oxydation du toluène cités ci-dessus sont listés dans le Tableau 4.

Nombre d'espèces	Nombre de réactions	Réacteur	Validations			Référence
			T [K]	P [atm]	ϕ [−]	
Non-précisé	130	FLR	1100-1200	1	0,65-1,33	Emdee et al. (1992)
659	109	JSR	1200	1	0.3-1.6	Lindstedt et al. (1996)
97	529	FLR	1170	1	0,76-1.25	Klotz et al. (1998)
65	340	FLR ST	1100-1200 1250-1550-	1 2,2-6,8	0,65-1,33 1,8	Djurisic et al. (2001)
379	<1000	FLR ST	1173 1300-1900	1 8-9,4	0,76 0,5-1,5	Pitz et al. (2001)
120	920	JSR ST	1000-1375 1300-1800	1 2,5	0,5-1,5 0,5-1,5	Dagaut et al. (2002)
97	538	ST	1250-1450	22-550	1 et 5	Sivaramakrishnan et al. (2004)
148	1036	FLR ST JSR	873-923 1200-1800 873-923	1 1,1-6,13 1	0,63-1,4 0,5-5 0,45-0,91	Bounaceur et al. (2005)

Tableau 4   Sélection de quelques mécanismes détaillés écrits pour l'oxydation du toluène.

3.2 Les modèles d'oxydation de mélanges binaires PRF (n-heptane/iso-octane)

Leppard (1992) a effectué une étude sur l'oxydation du mélange iso-octane/n-heptane dans un moteur CFR (Cooperative Fuel Research Engine). Il observe qu'en mélange PRF, le n-heptane est décomposé plus rapidement que l'iso-octane. Pour Leppard, la diminution de la réactivité avec l'augmentation de l'indice d'iso-octane s'explique par une décomposition plus rapide du n-heptane donnant des radicaux qui ont une probabilité égale de réagir avec le n-heptane et avec l'iso-octane : les radicaux réagissant avec l'alcane branché entrent dans un mécanisme de réaction en chaîne plus lent que dans le cas du n-heptane, ce qui explique, selon Leppard, la diminution de la réactivité. Ces résultats ont motivé un grand nombre d'études sur la chimie de l'oxydation des mélanges n-heptane/iso-octane.

Callahan et al. (1996) ont utilisé un regroupement des mécanismes publiés par l'équipe de Milan pour l'oxydation du n-heptane pur (Ranzi, 1995b) et de l'iso-octane pur (Ranzi, 1997). Leur mécanisme a été validé à partir d'expériences menées en réacteur piston adiabatique pour des températures allant de 550 à 1250 K, en mélange stoechiométrique. Le modèle reproduit bien les profils de température et de pression pour des mélanges de différents indices d'octane. Leur mécanisme a également été validé sur des résultats obtenus en machine à compression rapide. Le modèle reproduit correctement les délais d'auto-inflammation pour des mélanges d'indice d'octane 90 et 95.

Davis et Law (1998) ont proposé un mécanisme pour l'oxydation des PRF. En partant des mécanismes de Curran et al. (2002) pour l'iso-octane et de Held et al. (1997) pour le n-heptane, ils ont construit un mécanisme semi détaillé pour l'oxydation du mélange n-heptane/iso-octane contenant 69 espèces réagissant selon 403 réactions. Ce modèle a été utilisé par Huang et al. (2004) pour simuler des vitesses de flamme laminaire prémélangée.

Andrae et al. (2005) ont proposé un mécanisme décrivant l'oxydation des mélanges PRF et également celle des mélanges n-heptane/toluène. Pour cela ils ont rajouté 131 réactions de co-oxydation entre le n-heptane et l'iso-octane au modèle d'oxydation des mélanges PRF de Curran et al, (2002) et 12 réactions de co-oxydation entre le n-heptane et le toluène au modèle d'oxydation du toluène de Dagaut et al. (2002). Le nombre de réactions indiquées dans le Tableau 5 inclut donc le mécanisme d'oxydation du toluène et les réactions de co-oxydation entre le n-heptane et le toluène.

Glaude et al. (2002) ont généré automatiquement grâce au logiciel EXGAS un mécanisme d'oxydation du mélange iso-octane/n-heptane. Leur mécanisme comprend 647 espèces qui réagissent par 2386 réactions et a permis de modéliser des résultats obtenus par

Dagaut et al. (1994) en réacteur parfaitement agité. Les mélanges étudiés varient en indice d'octane entre 10 et 90. L'analyse de sensibilité effectuée à 650 K montre que les processus influençant fortement la réactivité globale du système sont les mêmes que pour l'oxydation des composés purs. Ainsi les métathèses avec le radical •OH consommant le n-heptane et les réactions d'isomérisation des radicaux iso-octyles et peroxy-iso-octyles sont les processus les plus importants du mécanisme primaire. Une autre observation des auteurs est qu'à basse température (~650 K) la réactivité du mélange décroît très rapidement lorsque l'indice d'octane augmente, alors qu'à plus haute température (T >850 K) cette variation est moins grande. Ce mécanisme a été amélioré par Buda et al. (2005) pour reproduire les délais d'auto-inflammation.

Miyoshi (2005) a développé le logiciel KUCRS pour générer de façon automatique des mécanismes cinétiques détaillés. Ogura et al. (2007) ont utilisé KUCRS pour générer un mécanisme qui décrit l'oxydation de mélanges PRF. En plus, ils ont couplé ce mécanisme avec des réactions d'oxydation de l'éthanol et de l'éthyl-tert-butyl-ether (ETBE).

Les mécanismes détaillés d'oxydation de mélanges PRF mentionnés ci-dessus sont listés dans le Tableau 5.

Tableau 5   Aperçu des mécanismes complexes écrits pour l'oxydation des mélanges PRF.

Nombre d'espèces	Nombre de réactions	Réacteur	Validations T [K]	P [atm]	ϕ [−]	Référence
150	3000	RCM FLR JSR	630-910 550-850 550-1100	12-17 12,5 10	1 1 1	Callahan et al. (1996)
69	403	Flamme prémelangée FLR	- 1080	1 1	0,6-1,8 1	Davis et Law (1998)
860	3600	validé pour l'iso-octane (voir tableau 3)				Curran et al. (2002)
647	2386	JSR	550-1000	1-10	0.2-1	Glaude et al. (2002)
351	1684	RCM ST	650-900 650-1250	12-17 12-50	1 1	Buda (2005)
1034	4238	ST RCM HCCI-moteur	650-1200 318	10 40-44	1 0,4 0,2-0.25	Andrae et al. (2005)
634	2930	ST FLR JSR CFR-moteur	700-1250 1100 600-1100	2, 40 1 1, 10	1 0.61 0.5-2	Ogura et al. (2007)

3.3 Les modèles d'oxydation de mélanges PRF/toluène

Naik et al. (2005a, 2005b) ont proposé un schéma cinétique pour l'oxydation d'un carburant modèle représentatif d'une essence composée de n-heptane, d'iso-octane, de 1-pentène, de toluène et de méthylcyclohexane. Ce modèle est basé sur les mécanismes de Curran pour l'oxydation des PRF (Curran et al. 2002) et du modèle d'oxydation du toluène de Pitz et al. (2001).

En 2007, Andrae et al. (2007) ont proposé un modèle pour les mélanges PRF/toluène. Ce modèle est basé sur le modèle d'Andrae et al. (2005) et contient 1083 espèces et 4535 réactions. Il reproduit correctement des résultats expérimentaux obtenus en tube à onde de choc pour l'oxydation de mélanges n-heptane/toluène et pour l'oxydation des mélanges PRF/toluène. Le modèle a également été validé par rapport à des expériences réalisées en moteur HCCI.

Un autre modèle pour des mélanges PRF/toluène a été proposé par Chaos et al. (2007). Les auteurs ont couplé les modèles de Curran et al. (2002) pour l'oxydation des mélanges PRF avec le modèle d'oxydation du toluène proposé par Klotz et al. (1998). Le modèle a été validé en réacteur tubulaire à basse température entre 850 et 950 K.

Récemment, Sakai et al. (2009) ont présenté un mécanisme d'oxydation des mélanges PRF/toluène en couplant une version améliorée du modèle d'oxydation de toluène de Pitz et al. (2001) avec celui pour l'oxydation des mélanges PRF en présence d'éthyl-tert-butyl éther (ETBE) et d'éthanol proposé par Ogura et al. (2007). Les mécanismes détaillés d'oxydation de mélanges PRF/toluène cités ci-dessus sont listés dans le Tableau 6.

Tableau 6   Sélection de des mécanismes détaillés écrits pour l'oxydation de mélanges PRF/toluène.

Nombre d'espèces	Nombre de réactions	Réacteur	Validations			Référence
			T [K]	P [atm]	ϕ [−]	
1328	5835	ST	850-1280	15-60	0.2-1	Naik et al.
		HCCI	-	-	0.08-0.32	(2005b)
1034	4238	ST	720-1100	10	1	Andrae
		RCM	318	-	0.4	(2005)
		HCCI-moteur		1-2	0.2-0.25	
1083	4535	ST	710-1300	30-50	0.3-1	Andrae et
		HCCI	-	1-2	0.2-0.25	al. (2007)
469	1221	FLR	850-950	12.5-55	0.3-1	Chaos et al.
		ST	830-1180	10-50	0.3-1	(2007)
		HCCI-moteur	-	-	0.286	
		Flamme prémelangée	-	1	0.6-1.4	
783	2883	ST	500-1700	2-50	0.25-1.2	Sakai et al.
		FLR	550-950	12.5	1	(2009)

3.4 Les modèles cinétiques des interactions NO$_x$-hydrocarbures

Parmi les modèles cinétiques proposés dans la littérature pour décrire les interactions entre NO et les hydrocarbures, nous avons choisi de présenter ici ceux de Moréac (2003) et de Glaude et al. (2005).

3.4.1. Le modèle de Moréac (2003)

Le modèle de Moréac (2003) est basé sur les mécanismes de l'oxydation du n-heptane et de l'iso-octane publiés par Curran et al. (1998, 2002) et sur le mécanisme détaillé de Heldt et al. (1994) proposé pour l'oxydation du méthanol. Les réactions de couplage entre les hydrocarbures et les espèces azotées écrites par Moréac sont inspirées d'un couplage proposé par Dagaut et al. (1999) pour décrire la combustion du gaz naturel en présence du NO ainsi que les effets de NO sur l'oxydation de l'éthane à basse température. Le sous-mécanisme pour la basse température est basé sur les schémas cinétiques développés par Bromly et al. (1992, 1996) complétés par des données cinétiques d'Atkinson et al. (1992) et de Wallington et al. (1992). Les différents mécanismes cités ci-dessus ont été couplés et le schéma cinétique ainsi établi s'est avéré adéquat pour le régime des températures intermédiaires. A basse température, pour les hydrocarbures avec une longue chaîne carbonée, comme le n-heptane et l'iso-octane, Moréac a introduit les réactions de couplage suivantes :

(33) $NO + ROO\bullet \leftrightarrow NO_2 + RO\bullet$

(35) $R^\bullet + NO_2 \leftrightarrow RNO_2$

(36) $RO^\bullet + NO_2 \leftrightarrow RONO_2$

Dans ce domaine de température, les radicaux alkylpéroxyles RO$_2\bullet$ sont formés en abondance lors de l'oxydation des hydrocarbures et peuvent donc interagir avec NO. Il a été montré que la suite des réactions suivantes déjà présentées dans le chapitre précédent :

(31) $NO + \bullet OOH \leftrightarrow NO_2 + \bullet OH$

(32) $NO_2 + H\bullet \leftrightarrow \bullet OH + NO$

mène à la formation rapide de radicaux •OH qui réagissent plus rapidement avec l'hydrocarbure de départ que les atomes •H, d'où une production accélérée de radicaux alkyles. Une autre voie d'inhibition de l'oxydation des alcanes à basse température a aussi été identifiées; la réaction de formation de l'acide nitreux :

(34) $NO + \bullet OH + M \leftrightarrow HONO + M$

où HONO est une espèce stable dont la formation entraîne une rupture de chaîne par consommation de radicaux •OH.

3.4.2. Le modèle d'oxydation de Glaude et al. (2005)

Glaude et al. (2005) ont écrit un mécanisme d'oxydation du n-pentane qui prend en compte les interactions des alcanes avec les NO$_x$. Leur mécanisme est basé sur un mécanisme d'oxydation des alcanes généré automatiquement par le logiciel EXGAS (Warth et al. 1998). Les réactions couplant les hydrocarbures avec les espèces azotées sont, d'une part, structurées dans une base des réactions des espèces en C_0-C_2 interagissant avec les espèces azotées et, d'autre part, concernent aussi des réactions de couplage pour les espèces possédant un nombre d'atomes de carbone supérieur à 2. Les réactions des espèces en C_0-C_2 avec les composées azotées sont inspirées de celles présentes dans le mécanisme GRI.2.11 et qui ont été publié par Atkinson et al. (1992). Dans le mécanisme de Glaude et al. (2005), les NO$_x$ interagissent avec les radicaux péroxyalkyles ROO• et les hydropéroxypéroxyalkyles (•OOQOH) via réactions :

(33)   NO + ROO• ↔ NO$_2$ + RO•

(37)   •OOQOOH + NO ↔ NO$_2$ + •OH + produits

Glaude et al. (2005) ont défini deux voies compétitives pour la décomposition des radicaux RO•. Une première voie (voie réactive A) consiste en une isomérisation des radicaux RO• et une nouvelle addition d'oxygène sur les radicaux résultants •ROH :

(38)   RO• ↔ •ROH

(39)   •ROH + O$_2$ ↔ •OOQOH

Le radical formé •OOQOH s'isomérise en produisant un radical HOO•Q'OH :

(40)   •OOQOH ↔ HOO•Q'OH

L'isomérisation des radicaux •OOQOH en HOO•Q'OH est en compétition avec la réaction suivante faisant intervenir NO :

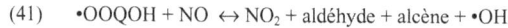

(41)   •OOQOH + NO ↔ NO$_2$ + aldéhyde + alcène + •OH

Les radicaux HOOQ•OH peuvent aussi se décomposer en éthers cycliques ou réagir encore par addition d'oxygène :

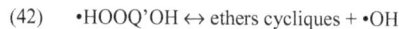

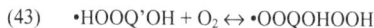

(42)   •HOOQ'OH ↔ ethers cycliques + •OH

(43)   •HOOQ'OH + O$_2$ ↔ •OOQOHOOH

Les radicaux $\bullet OOQOHOOH$ sont enfin consommés par la réaction suivante:

$$(44) \quad \bullet OOQOHOOH \leftrightarrow HCHO + \bullet OH + \bullet OOH + \text{aldéhyde} + \text{oléfine}$$

La deuxième voie de décomposition (voie réactive B) des radicaux $RO\bullet$ proposée par Glaude consiste en une décomposition directe des $RO\bullet$ par béta-scission par le type de réaction suivant:

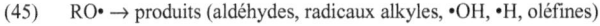

$$(45) \quad RO\bullet \rightarrow \text{produits (aldéhydes, radicaux alkyles, } \bullet OH, \bullet H, \text{ oléfines)}$$

Selon Glaude et al. (2005), les NO_x interagissent aussi avec les radicaux allyliques, stabilisés par résonance. Les interactions entre les NO_x et ces espèces sont définies par les réactions du type suivant :

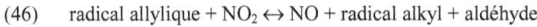

$$(46) \quad \text{radical allylique} + NO_2 \leftrightarrow NO + \text{radical alkyl} + \text{aldéhyde}$$

La Figure 3-1 présente le schéma des réactions d'interaction entre les hydrocarbures et les NO_x telles que définies dans le mécanisme de Glaude et al. (2005).

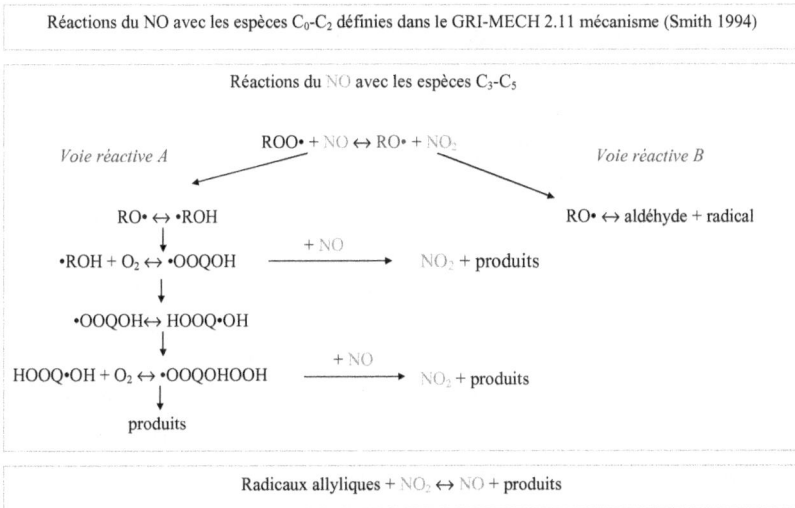

Figure 3-1   Diagramme résumant les réactions d'interaction entre les NO_x et les hydrocarbures selon Glaude et al. (2005).

Le mécanisme proposé par Moréac a été développé à partir de schémas d'oxydation des hydrocarbures élaborés dans différents laboratoires et écrits de manière non générique, ce qui augmente sa complexité et réduit sa généralité. Le mécanisme proposé par Glaude et al. (2005)

a été généré automatiquement, les réactions rajoutées étant limitées uniquement à celles couplant les espèces azotées avec les hydrocarbures et l'oxygène. Les réactions d'oxydation des alcanes, générées automatiquement, n'ont pas été modifiées. Le mécanisme proposé par Glaude et al. (2005) possède ainsi une structure bien définie qui facilite son adaptation au cas des hydrocarbures d'une taille supérieure au n-pentane.

3.5 Conclusions sur l'état de l'art des modèles d'oxydation des hydrocarbures

Deux types de mécanismes détaillés pour l'oxydation des hydrocarbures sont proposés dans la littérature : les mécanismes générés automatiquement en appliquant des règles cinétiques génériques et ceux pour lesquels chaque réaction est écrite après une recherche bibliographique. Le principal avantage des mécanismes générés automatiquement est lié à l'application systématique de règles génériques, ce qui limite les risques d'erreur et d'oubli et permet de traiter aisément toute une famille de réactifs. Les mécanismes écrits manuellement peuvent néanmoins parfois permette une meilleure prédiction dans le cas d'une espèces particulière.

Jusqu'à la fin des années 90, les efforts de modélisation ont été concentrés sur la modélisation de l'oxydation des hydrocarbures simples (n-heptane, iso-octane ou toluène). Par la suite, un large nombre de mécanismes ont été proposés pour modéliser l'oxydation de mélanges binaires, notamment des mélanges PRF. Dans le but de proposer des carburants modèles plus représentatifs de carburants réels contenant des espèces aromatiques, un certain nombre de mécanismes a été écrit pour des mélanges ternaires contenant également du toluène. Quand les travaux présentés dans ce mémoire ont commencé, plusieurs mécanismes décrivant l'oxydation de mélanges PRF/toluène étaient déjà disponibles, en revanche aucun de ces modèles n'était basé sur des mécanismes générés automatiquement, ni ne tenait compte de l'effet de NO sur l'oxydation des hydrocarbures. Aucun d'entre eux non plus n'avait été validé sur une très large gamme de température et pression.

Chapitre 4

Kinetic modelling of PRF/toluene oxidation in presence of NO$_x$

The literature review presented in Chapter 3 revealed that the modelling of postoxidation phenomena in engine exhaust gases has to take into account complex chemistry. Complex chemistry is generally modelled by detailed reaction mechanisms. In the case of postoxidation, these have to consider low temperature chemistry as well as the impact of the different species present in the burned gases. However, the number of different species in the engine exhaust gases is much larger than the number that can be taken into account in detailed reaction mechanisms. For the kinetic model developed in this work, we have chosen as « key-species »: n-heptane, iso-octane, toluene and NO (see Section 1.4). In the following, we describe the construction of a detailed mechanism for the oxidation of n-heptane, iso-octane and toluene mixtures and the coupling of those reaction mechanisms. One major part of this work was the development of a low temperature mechanism for the interaction of NO with hydrocarbons. The choices of the different reactions as well as the respective kinetic constants is presented and justified. Thorough validation of the whole kinetic mechanism has been performed against a large number of experimental results in different configurations under a wide range of thermodynamic conditions and is also presented in this section. The work described in this section has been the subject of a published paper in the journal Combustion and Flame:

(Anderlohr J.M., Bounaceur R., Pires da Cruz A., Battin-Leclerc F. (2009), Modelling of Autoignition and NO Sensitization for the Oxidation of IC-Engine Surrogate Fuels, Combust. Flame, 156, 505-521)

4.1 Description of the proposed detailed kinetic mechanism

The mechanism that we propose here is based on the PRF-autoignition model from Buda et al. (2005) coupled with the model for the oxidation of toluene provided by Bounaceur et al. (2005). Additionally; the reactions of NO$_x$ with PRF and toluene compounds were written. For reactions of small species, pressure-dependent rate constants are defined by the formalism proposed by Troe et al. (1974). Thermochemical data for molecules and radicals not tabulated were calculated by the THERGAS software (Muller et al. 1995), which is based on additivity methods proposed by Benson et al. (1976). It yields 14 polynomial coefficients according to the CHEMKIN formalism (Kee et al. 1993). In the case of nitrogen containing compounds, thermochemical data proposed by Marinov et al. (1998) and

Burcat and Rusic (2007) have been used. The new mechanism contains 3000 reactions and 536 species.

4.1.1. Mechanism for the oxidation of n-heptane and iso-octane

A mechanism for the oxidation of a n-heptane/iso-octane mixture has been generated by the software EXGAS according to the principles described by Buda et al. (2005) and Warth et al. (1998).

4.1.1.1. Description of the EXGAS-software

The EXGAS software generates automatically detailed reaction mechanisms for the oxidation of alkanes. The EXGAS-mechanisms are generated by well defined rules for alkane decomposition. These are coupled with data from separate kinetic and thermodynamic generators. All generated mechanisms are written in the CHEMKIN-format. Figure 4-1 shows schematically the described architecture of the EXGAS software.

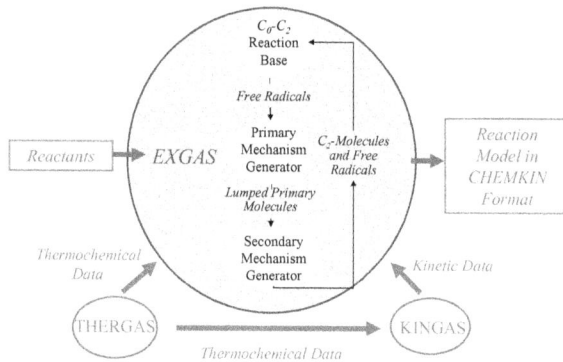

Figure 4-1   Principle scheme of EXGAS generator (Warth et al. 1998).

The EXGAS mechanisms as well as the mechanism presented here are composed of three parts:

- A C_0-C_2 reaction base (Barbé et al. 1995) involving species with up to two carbon atoms and including kinetic data mainly taken from the evaluations of Baulch et al. (1994) and Tsang and Hampson (1986). To obtain a good agreement with the experimental results obtained at high pressures (above 40 atm), the rate constant of the decomposition of H_2O_2 defined in the C_0-C_2 reaction base had to be multiplied by a factor 4, compared to the rate constant defined for lower pressures. The need for this adjustment is probably related to a still unknown problem in the mechanism at high pressure.

- A comprehensive primary mechanism. It only considers initial organic compounds and oxygen as reactants. It includes all the usual low and intermediate temperature reactions of alkanes, which are:
  - o Unimolecular initiations involving the breaking of a C-C bond.
  - o Bimolecular initiations with O$_2$ to produce alkyl (R•) and hydroperoxy (•OOH) radicals.
  - o Additions of alkyl and hydroperoxyalkyl (•QOOH) radicals to an oxygen molecule.
  - o Isomerizations of alkylperoxy and hydroperoxyperoxy radicals (ROO• and •OOQOOH) involving a cyclic transition state, we consider a direct isomerization-decomposition to give ketohydroperoxides and hydroxyl radicals (Glaude et al. 2000).
  - o Decompositions of radicals by β-scission involving the breaking of C-C or C-O bonds for all types of radicals (for low temperature modelling, the breaking of C-H bonds is not considered).
  - o Decompositions of hydroperoxyalkyl radicals to form cyclic ethers and •OH radicals.
  - o Oxidations of alkyl radicals with O$_2$ to form alkenes and •OOH radicals.
  - o Metatheses between radicals and the initial reactants (H-abstractions).
  - o Recombinations of radicals.
  - o Disproportionations of peroxyalkyl radicals with •OOH to produce hydroperoxides and O$_2$.
- A lumped secondary mechanism. The molecules produced in the primary mechanism, with the same molecular formula and the same functional groups are lumped into one unique species without distinction between different isomers. The secondary mechanism includes global reactions producing, in the smallest number of steps, molecules or radicals whose reactions are included in the C_0-C_2 base.

Previous work on PRF fuels shows that co-oxidation reactions between n-heptane and iso-octane are negligible and that the coupling between their oxidation kinetics is based mainly on small radicals interactions (Glaude et al. 2002) defined in the C_0-C_2 base. For the coupling between PRF fuels and toluene we wrote cross-reactions between n-heptane/iso-octane and benzyl and peroxyalkyl radicals, as well as the cross-reactions between allylic radicals and toluene.

4.1.1.2. Description of the generic rate constants

Table 7 presents the set of generic rate constants, which are used by EXGAS for the primary mechanism of the oxidation of alkanes, apart from the kinetic data of isomerizations, recombinations and the unimolecular decompositions, which are calculated using software KINGAS (Bloch-Michel, 2005), based on thermochemical kinetics methods (Benson, 1976).

Table 7   *Kinetic parameters for the primary mechanism of the oxidation of alkanes not calculated using KINGAS (Bloch-Michel, 2005).*

a) H-abstraction

with :	Primary H			Secondary H			Tertiary H		
	$\lg A$	b	E_a	$\lg A$	b	E_a	$\lg A$	b	E_a
Initiation with O_2	12.84	0	THERGAS	12.84	0	THERGAS	12.84	0	THERGAS
Oxidation : n>4	11.43	0	5000	11.99	0	5000	11.80	0	5000
Oxidation : n≤4	11.60	0	5000	12.16	0	5000	11.99	0	5000

b) Metathesis

with :	$\lg A$	b	E_a	$\lg A$	b	E_a	$\lg A$	b	E_a
•O•	13.23	0	7850	13.11	0	5200	13.00	0	3280
•H	6.98	2	7700	6.65	2	5000	6.62	2	2400
•OH	5.95	2	450	6.11	2	-770	6.06	2	-1870
•CH₃	-1	4	8200	11.0	0	9600	11.00	0	7900
HO₂•	11.30	0	17000	11.30	0	15500	12.00	0	14000
•CHO	4.53	2.5	18500	6.73	1.9	17000	4.53	2.5	12000
•CH₂OH	1.52	2.95	14000	1.48	2.95	12000	2.08	2.76	10800
•OCH₃	10.73	0	7300	10.86	0	45000	10.36	0	2900
•OOR	12.30	0	20000	12.18	0	17500	12.18	0	15000
•C₂H₅	11.00	0	12000	11.00	0	11000	11.00	0	9200
i-C₃H₇•	-2.85	4.2	8700	-2.85	4.2	8000	-2.85	4.2	6000
•Rₚ	11.00	0	12000	11.00	0	11200	11.00	0	9000
•Rₛ	11.00	0	14500	11.00	0	12200	11.00	0	10000
•Rₜ	11.00	0	15000	11.00	0	12700	11.00	0	10500

c) Other reactions

		$\lg A$	b	E_a
Addition of an alkyl radical to O_2		See text		
C—C breaking	•CH₃ + molecule	13.30	0	31000
	• Rₚ + molecule	13.30	0	28700
	• Rₛ + molecule	13.30	0	27700
	• Rₜ + molecule	13.30	0	26700
C—O breaking	HO₂• + molecule	12.92	0	26000
O—O breaking	•OH + molecule	9.00	0	7500
Formation of cyclic ethers	cycle with 3 atoms	11.78	0	17950
	cycle with 4 atoms	10.96	0	16600
	cycle with 5 atoms	9.56	0	7000
	cycle with 6 atoms	8.23	0	1950
•OOR and HO₂• disproportionation		11.30	0	-1300

Note 1: Rate constants are expressed in the form $k = A\,T^b \exp(-E/RT)$, with the units cm³, mol, s, kcal, by H atoms which can be abstracted. •Rₚ, •Rₛ, •Rₜ are primary, secondary and tertiary alkyl free radicals. n is the number of atoms of carbon in an alkyl radical.

Note 2: In case of the oxidation of iso-octyl radicals the rate constants have been devided by 3.

Rate constants of the following reaction types are automatically generated:

- *Addition of branched alkyl radical to an oxygen molecule*

The rate constant of the addition of alkyl radical to an oxygen molecule has an important impact, as it directly competes with the oxidation reactions. For linear and branched alkyl radicals, we use an additivity method to take into account the structure of each considered radical:

$$[1] \quad k_{add} = n_p.k_p + n_s.k_s + n_t.k_t + n_q.k_q$$

With:

n_p = *number of primary groups (CH_3) linked to the radicalar atom of carbon*

n_s = *number of secondary groups (CH_2) linked to the radicalar atom of carbon*

n_t = *number of tertiary groups (CH) linked to the radicalar atom of carbon*

n_q = *number of quaternary groups (C) linked to the radicalar atom of carbon*

and coefficients k as follows:

$$k_p = 8.0 \times 10^{18} \, T^{-2.5} \, cm^3.mol^{-1}.s^{-1}$$

$$k_s = 9.0 \times 10^{18} \, T^{-2.5} \, cm^3.mol^{-1}.s^{-1}$$

$$k_t = 1.5 \times 10^{18} \, T^{-2.5} \, cm^3.mol^{-1}.s^{-1}$$

$$k_q = 1.0 \times 10^{18} \, T^{-2.5} \, cm^3.mol^{-1}.s^{-1}$$

The principle of this additivity method was proposed by Stocker and Pilling (1997). For a tertiary radical, $k_{add} = k_t$ if the atom of carbon carrying the radical centre is linked to at least one tertiary group and $k_{add} = k_q$ if it is linked to at least one quaternary group. Table 8 presents an example for the applied additivity method.

Table 8   Kinetic parameters for the additions of iso-octyl radicals to an oxygen molecule.

Considered radical	k_{add} [$cm^3.mol^{-1}.s^{-1}$]
	$k = k_t = 1.5 \times 10^{18}.T^{-2.5}$
	$k = k_s + 2k_p = 2.5 \times 10^{19}.T^{-2.5}$
	$k = k_t + k_q = 2.5 \times 10^{18}.T^{-2.5}$
	$k = k_q = 1.0.10^{18} \times T^{-2.5}$

- *Isomerizations of alkylperoxy radicals to hydroperoxy alkyl radicals*

As described by Warth et al. (1998), pre-exponential factors (A) are mainly based on the changes in the number of internal rotations as the reactant moves to the cyclic transition state.

That correlation is described by equation [2].

$$[2] \quad A = \frac{2.718 \cdot k_B T}{h_p}(rpd) \cdot \exp^{\left(\frac{3.5(\Delta n_{int.rot}+1)}{R}\right)}$$

with $\Delta n_{int.rot}$ equal to the change in the number of internal rotations as the reactant moves to the transition state, h_p equal to the Planck constant, k_B equal to the Boltzmann constant, R as the gas constant, rpd equal to the number of identical abstractable H atoms and T the temperature. Activation energies have been set equal to the sum of the activation energy for H-abstraction from the substrate by analogous radicals ($E_{H\text{-}abstr}$, see Table 9a) and the strain energy of the cyclic transition state (E_{ring}, see Table 9b) and are calculated as follows.

$$[3] \quad E = E_{H\text{-}abstr} + E_{ring}$$

The activation energy for H-abstractions $E_{H\text{-}abstr}$ of alkylperoxy radicals is estimated particulary for the cases (1) and (-1) illustrated in Figure 4-2.

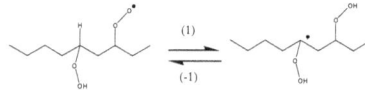

Figure 4-2   Isomerizaton of alkylperoxy radicals (Buda, 2005).

In the case of isomerization (1), $E_{H\text{-}abstr}$ is considered to be 2 kcal/mol lower than those of the similar reactions when a carbon atom is linked to only carbon atoms, hydrogen atoms or a radical centre. The value $(\Delta E_1 = 2$ cal/mol) corresponds to the difference in activation energy between the abstraction by ethyl radicals at 800 K of an atom of hydrogen from methanol (Tsang et al. 1987) and that of a primary atom of hydrogen from propane, (Tsang et al. 1987).

Table 9 presents activation energies for isomerizations of alkylperoxy and hydroperoxyperoxy radicals and the strain energies of cyclic compounds containing two O-atoms.

Table 9   Parameters used for the calculation of the rate constants of internal isomerizations of alkylperoxy and hydroperoxy alkylperoxy free radicals.

		Primary -CH$_3$	Secondary-CH$_2$	Tertiary -CH
a)	Activation energy $E_{H\text{-}abstr}$ for an H-atom abstraction by ROO•	$\left[kcal/mol\right]$	$\left[kcal/mol\right]$	$\left[kcal/mol\right]$
		20	17	14

		Size of the ring (number of C-atoms n_{Carb})				
b)	Strain energy E_{ring} of cyclic compounds containing two O atoms n_{Carb} [-]	4	5	6	7	8
	E_{ring} $\left[kcal/mol\right]$	23	15.5	8.0	5.0	4.0

Table 10 presents examples of rate constants obtained for isomerizations and details the calculation of the activation energies including a case for which the 2 kcal/mol correction is needed.

Table 10   *Examples of rate parameters in the case of the oxidation of n-butane.*

	A [s^{-1}] [a]	E_a [kcal.mol^{-1}] [a]
From butyl-1-peroxy radicals	$3.3 \times 10^9 \times T$	$E_a = 17 + 15.5 = 32.5$
	$5.7 \times 10^8 \times T$	$E_a = 17 + 8 = 25$
	$1.5 \times 10^8 \times T$	$E_a = 20 + 5 = 25$
From 3-hydroperoxy butyl-1-peroxy radicals [b]	$3.3 \times 10^9 \times T$	$E_a = 17 + 15.5 = 32.5$
	$2.9 \times 10^8 \times T$	$E_a = 14 + 8 - 2 = 20$
	$1.5 \times 10^8 \times T$	$E_a = 20 + 5 = 25$

[a] The reactions are considered as reversible and the rate parameters are those of the direct step.
[b] The direct isomerization/decomposition is considered with the rate constant of the isomerization.

- *Recombination of free radicals and unimolecular decomposition of molecules*

The rate constants for unimolecular decompositions are obtained from the data of the corresponding reverse reaction and the thermochemical data calculated by THERGAS. Activation energies for recombination reactions are taken to be zero (Warth et al. 1998). For the recombination of two free radicals •R and •R' to form the molecule R-R', the pre-exponential factor is estimated from modified collision theory to be:

$$[4] \quad A = \frac{1}{4} \cdot \left(2.8 \cdot l_{R-R} \cdot \left[\frac{300}{T} \right]^{0.17} \right)^2 \left(\frac{8\pi RT}{\mu} \right)^{1/2} \cdot P$$

With $l_{R-R'}$ the length of the R-R' bond, μ the reduced molar mass and P an empirically estimated, steric factor.

4.1.2. Mechanism for the oxidation of toluene

The model for the oxidation of toluene is that of Bounaceur et al. (2005) and includes the following sub-mechanisms:

- A primary mechanism including reactions of toluene containing 193 reactions and including the reactions of toluene and of benzyl, tolyl, peroxybenzyl (methylphenyl), alkoxybenzyl and cresoxy free radicals.

- A secondary mechanism for the oxidation of toluene involving the reactions of benzaldehyde, benzyl hydroperoxide, cresol, benzylalcohol, ethylbenzene, styrene and bibenzyl.

- A mechanism for the oxidation of benzene (Da Costa et al. 2003). It includes the reactions of benzene and of cyclohexadienyl, phenyl, phenylperoxy, phenoxy, hydroxyphenoxy, cyclopentadienyl, cyclopentadienoxy and hydroxycyclopentadienyl free radicals, as well as the reactions of ortho-benzoquinone, phenol, cyclopentadiene, cyclopentadienone and vinylketene.

- A mechanism for the oxidation of unsaturated C_0-C_4 species. It contains reactions involving •C$_3$H$_2$, •C$_3$H$_3$, C$_3$H$_4$ (allene and propyne), •C$_3$H$_5$ (three isomers), C$_3$H$_6$, C$_4$H$_2$, •C$_4$H$_3$ (2 isomers), C$_4$H$_4$, •C$_4$H$_5$ (5 isomers), C$_4$H$_6$ (1,3-butadiene, 1,2-butadiene, methyl-cyclopropene, 1-butyne and 2-butyne).

In case of redundant reactions defined in both the secondary mechanism generated for the alkanes oxidation and in the C_0-C_4 reaction base for the toluene oxidation (e.g. reaction of propene or of allyl radicals), we kept the reactions written in the n-heptane/iso-octane mechanism and skipped those defined in the toluene mechanism. The mechanism for the oxidation of toluene has been validated in previous work (Bounaceur et al. 2005) using experimental results obtained in JSR, plug flow reactors and shock tubes.

4.1.3. Cross-term reactions between alkanes and toluene

The following co-oxidation reactions between alkanes and toluene are defined:

- Metathesis of benzyl radicals with n-heptane and iso-octane leading to toluene and alkyl radicals (respectively heptyl and octyl) with an A factor of $1.6 \times T^{3.3}$ cm^{3}mol^{-1}s^{-1} and an activation energy of 19.8, 18.2 and 17.2 kcal/mol for the abstraction of a primary, secondary and tertiary H-atom, respectively (Heyberger et al. 2002).

- Terminations between benzyl and alkyl radicals producing alkylbenzenes with a rate constant of 1.0×10^{13} cm^{3}mol^{-1}s^{-1} according to collision theory.

- Metathesis with toluene of secondary allylic radicals (iso-butenyl, iso-octenyl, heptenyl) with a rate constant of 1.6×10^{12}exp(-7600/T) cm^{3}mol^{-1}s^{-1} as for allyl radicals (Bounaceur et al. 2005) and of methyl peroxy (CH$_3$OO•) radicals with a rate constant of 4.0×10^{13}exp(-6040/T) cm^{3}mol^{-1}s^{-1} (Andrae et al. 2005).

4.1.4. Reactions of NO$_x$ compounds

The reaction mechanism for the NO$_x$ species is shown in appendix A.1. It was derived from the modelling work of Glaude et al. (2005) concerning the effect of the addition of NO on the oxidation of n-butane and n-pentane. This mechanism was based on the model of Hori et al. (1998) on the conversion of NO to NO$_2$ promoted by methane, ethane, ethylene, propane, and propene. The kinetic model proposed by Glaude is primarily based on the GRI-MECH 2.11 (Bowmann et al. 1997) and on the research performed by Dean and Bozzelli (1997) and Atkinson et al. (1992). In this study, we used the more recent mechanism GRI-MECH 3.0 (Smith et al. 1999). Table 11 shows the reactions of NO$_x$ compounds which have been changed or added compared to the mechanisms of Glaude et al. (2005) and to the GRI-MECH 3.0 (Smith et al. 1999).

The rate parameters of reactions (34, 49, 49-51, 53-57) were chosen from recent studies on reaction kinetics between C_1-C_2 species with nitrogen containing compounds (Atkinson et al. 2005;   Choi et al. 2005;   Srinivasan et al. 2005;   Atkinson et al. 2002; Xu et al. 2003; Mebel et al. 1998; Diau et al. 1995; Tsang et al. 1991; He et al. 1988) and are different from those used by Glaude et al. (2005) for the same reactions. Rate parameters of reactions (50) and (54) have been adjusted in order to obtain satisfactory simulation results. The reactions of NO$_x$ with formaldehyde (HCHO) (reactions 50 and 51) and with nitrous acid (HONO) (reaction 57) and six reactions involving HNO radicals (reactions 52-56) were added with rate constants mainly taken from the literature.

Table 11   *Modified or added reactions of NO$_x$ compounds (*compared to Glaude et al. (2005) or GRI-MECH 3.0 (Simth et al. 1999)) *and cross-term reactions involved in the oxidation of alkanes by the addition of NO.*

Reactions		A	n	E_a	References	No
NO+•OH=HONO	(high pressure)	1.1×10^{14}	-0.3	0.0	Atkinson et al. (2004)	(34)
	(low pressure)	2.35×10^{23}	-2.4	0.0		
	Fall off :	F_c =0.81				
•CH$_3$+NO$_2$=CH$_3$O•+NO		1.36×10^{13}	0.0	0.0	Srinivasan et al. (2005)	(47)
CH$_3$NO$_2$=•CH$_3$+NO$_2$	(high pressure)	1.8×10^{17}	0.0	58500	Estimated[a]	(48)
	(low pressure)	1.3×10^{18}	0.0	42000		
CH$_3$O•+NO=HCHO+HNO		7.6×10^{13}	-0.76	0.0	Atkinson et al. (2005)	(49)
HCHO+NO=•CHO+HNO		1.02×10^{13}	0.0	40670	Tsang et al. (1991)	(50)
HCHO +NO$_2$=•CHO+HONO		8.35×10^{-11}	6.68	8300	Xu et al. (2003)	(51)
HNO+O$_2$=•OOH+NO		8.0×10^{10}	0.0	9520	Estimated[b]	(52)
HNO+NO$_2$=HNO$_2$+NO		3.0×10^{11}	0.0	1988	Tsang et al. (1991)	(53)
HNO+NO=•OH+N$_2$O		8.5×10^{12}	0.0	29640	Diau et al. (1995)	(54)
HNO+•CH$_3$=CH$_4$+NO		1.47×10^{11}	0.76	349.0	Choi et al. (2005)	(55)
HNO+CH$_3$O=CH$_3$OH+NO		3.16×10^{13}	0.0	0.0	He et al. (1988)	(56)
HONO+NO=NO$_2$+HNO		4.40×10^{03}	2.64	4038	Mebel et al. (1998)	(57)
C$_2$H$_5$• +NO$_2$=C$_2$H$_5$O•+NO		1.36×10^{13}	0.0	0.0	Estimated[c]	(58)
RH+NO$_2$=R•+HONO		$\alpha\times2.2\times10^{13}$	0.0	31100	Chan et al. (2001)[d]	(59)
		$\beta\times5.8\times10^{1}_{2}$	0.0	28100		
		$\gamma\times9.3\times10^{1}_{3}$	0.0	25800		
RNO$_2$= R•+NO$_2$	(high pressure)	1.8×10^{17}	0.0	58500	Estimated[e]	(60)
	(low pressure)	1.3×10^{18}	0.0	42000		
R•+NO$_2$ => RO•+NO		4.0×10^{13}	-0.2	0.0	Estimated[f]	(61)
ROO•+NO => RO•+NO$_2$		4.70×10^{12}	0.0	-358.0	Estimated[g]	(33)
•OOQOOH+NO=>NO$_2$+•OH+*HCHO+olefin[h]		4.70×10^{12}	0.0	-358.0	Estimated[h]	(37)
RO• => aldehyde+ R'•		2.0×10^{13}	0.0	15000	Curran et al. (1998)	(62)
aldehyde+NO$_2$=> R•+CO+HONO		8.35×10^{-11}	6.68	8300	Estimated[i]	(63)
Y•+NO$_2$ => acrolein + R'•+NO[b]		2.35×10^{13}	0.0	0.0	Glaude et al. (2005)	(64)

Left margin labels: *Oxidation of alkanes (C < 3)* ; *Oxidation of alkanes (C > 3)*

The rate constants are given (k=A T^n exp(-E$_a$/RT)) in cc, mol, s, cal units.
[a]: The rate constant was taken as 10 times the value proposed by Glänzer and Troe (1972).
[b]: The rate constant was taken as 3.6 times the value proposed by Bruykov et al. (1993) for the reaction between HNO and the O$_2$ radical.
[c]: The rate constant was taken as equal to that of reaction (49).
[d]: α is the number of primary H-atoms, β of secondary H-atoms and γ of tertiary H-atoms.
[e]: The rate constant was taken as equal to that of reaction (50).
[f]: The rate constant was taken similar to those proposed by Glarborg et al. (1998) for the reaction CH$_3$+NO$_2$=CH$_3$O•+NO.
[g]: The rate constant was taken as 1.8 times the value proposed by Atkinson et al. (1992) for the reaction between CH$_3$OO• radicals and NO.
[h]: The •OOQOOH decomposition defined as : CnH$_{(2n)}$OOOOH+NO => NO$_2$+OH +2*HCHO+C$_{(n-2)}$H$_{(2n-4)}$.
[i]: The rate constant was taken as equal to that of reaction (53).

Coupling reactions between species involved in the alkane oxidation model and NO$_x$ were written. These reactions were partly derived from the mechanism published by Glaude et al. (2005) for the oxidation of n-butane and n-pentane in presence of NO. The reaction types are:

- H-abstractions from alkanes by NO$_2$ (reaction 59) with rate parameters proposed by Chan et al. (2001).

- The reaction of alkyl radicals (R•) with NO$_2$ giving either RNO$_2$ (reaction 60) or alkoxy radicals (RO•) radicals (reaction 61) as it is written in the case of methyl radicals (•CH$_3$) radicals (reactions 47 and 48). The rate constant of reaction 65 was considered similar to that proposed by Glarborg et al. (1999) for the corresponding reaction of •CH$_3$ radicals with NO$_2$.

- Reactions of ROO• radicals with NO forming NO$_2$ and RO• (reaction (33)).

- Reactions of HOOQOO• radicals with NO which are decomposed via a global reaction producing NO$_2$, a hydroxyl radical (•OH), two HCHO molecules and the corresponding olefin (reaction (37)).

    The rate constants of reactions (33) and (37) were considered equal to the values proposed by Atkinson et al. (1992) for the corresponding reaction of methylperoxy radicals (CH$_3$OO•) with NO.

- Decomposition of RO• radicals by beta-scission (reaction 62) with the rate constant proposed by Curran et al. (1998).

- H-abstractions from aldehydes by NO$_2$ (reaction 63) with rate parameters estimated as in the case of formaldehyde (reaction 51).

- Reactions of resonance stabilized allylic radicals with NO$_2$ yielding NO, acrolein and an alkyl radical (R•) (reaction 64) with the rate constant proposed by Glaude et al. (2005).

Other reactions between NO$_x$ and hydrocarbons could have been taken into account but sensitivity analysis allowed us to neglect other reactions of alkoxy radicals, which were considered by Glaude et al. (2005). In contrast, the reactions of alkane molecules (reaction 14) and alkyl radicals (reactions 60) with NO$_2$ were added, while they were not considered by Glaude et al. (2005).

Our work for the modelling NO$_x$/alcane interactions was focused on some rate governing reactions. The alkane/NO$_x$ interactions found to be rate determining are listed in Table 12.

Table 12   *Rate governing reactions of nitrogen compounds for the oxidation of n-heptane and iso-octane.*

Reactions		A	n	E_a	References	No
ROO•+NO => RO•+NO$_2$		4.70x10^{12}	0.0	-358	Estimated[a]	(33)
CH$_3$OO•+ NO = CH$_3$O•+ NO$_2$		2.53x10^{12}	0.0	-358	Atkinson et al. (1992)	(65)
CH$_3$NO$_2$=•CH$_3$+NO$_2$	*(high pressure)*	1.8x10^{17}	0.0	58500	*10 x value proposed by Glänzer and Troe (1972).*	(48)
	(low pressure)	1.3x10^{18}	0.0	42000		
NO+•OOH=NO$_2$+•OH		2.11x10^{12}	0.0	-480	GRI-MECH 3.0	(31)
NO+•OH=HONO	*(high pressure)*	1.1x10^{14}	-0.3	0.0	Atkinson et al. (2004)	(34)
	(low pressure)	2.35x10^{23}	-2.4	0.0		

The rate constants are given (k=A T^n exp(-E$_a$/RT)) in cc, mol, s, cal units.
[a]: The rate constant was taken as 1.8 times the value proposed by Atkinson et al. (1992) for the reaction between CH$_3$OO• radicals and NO.

Interactions between peroxy radicals and NO play a major role for hydrocarbon oxidation kinetics at temperatures below 900 K. The reaction of ROO• radicals with NO producing NO$_2$ (reaction (33)) has an important impact on hydrocarbon oxidation. This reaction is in direct competition with the isomerization of peroxy radicals ROO• via reaction (7) which is followed by the second addition of oxygen by reaction (9).

(33)   ROO• + NO ↔ RO• + NO$_2$

(7)   ROO• ↔ •QOOH

(9)   QOOH + O$_2$ ↔ •OOQOOH

The hydroperoxyperoxy radicals •OOQOOH produced by reaction (9) decompose immediately to highly reactive OH radicals. Reaction (33) has thus a strong inhibiting effect on the global reactivity and its rate constant have to be chosen with care. In the context of peroxy radical / NO interactions the reaction between peroxymethyl radicals and NO is an exception. Methylperoxy radicals may not isomerise via reaction type (7). Such species include one carbon atom only and for that reason methylperoxy radicals are quite unreactive radicals. Methylperoxy radicals are transformed into methoxy radicals through reaction (65). These radicals decompose easily to produce reactive formaldehyde (HCHO) and •H atoms via reaction (66):

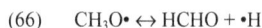

(65)   CH$_3$OO•+ NO ↔ CH$_3$O•+ NO$_2$

(66)   CH$_3$O• ↔ HCHO + •H

Reaction (65) transforms the less reactive CH$_3$OO• radicals into more reactive CH$_3$O• radicals. For that reason, reaction (65) has a very strong accelerating impact on global reactivity.

As a consequence, in presence of NO, the production of reactive CH$_3$O• is almost exclusively controlled by the formation of CH$_3$OO• radicals. CH$_3$OO• radicals are mainly produced from methyl radicals by addition to oxygen and thus methyl radicals are the major source of methylperoxy radicals.

(67)   CH$_3$•+ O$_2$ ↔ CH$_3$OO•

At low temperatures, the recombination of NO$_2$ with methyl radicals via the reverse of reaction (48) represents a sink of methyl radicals and reaction (48) controls the accelerating impact of NO via reaction (65) especially in presence of high concentrations of NO.

(48)   CH$_3$NO$_2$ ↔ •CH$_3$+NO$_2$

In order to control the strong accelerating impact of reaction (65), we have used a rather high rate constant of the recombination for methyl radicals with NO$_2$ (reaction (48)) compared to the rate constants proposed in literature. However, there might be uncertainties in collision efficiencies and in the thermodynamic data as well. Collision efficiencies of third body collision partners and thermodynamic properties of reaction partners/products are determining for the effective reaction rate of 3rd-body reactions and should be further investigated especially in the case of reaction (48). We confirmed the observations of Dubreuil et al. (2006) and Glaude et al. (2005) and Moréac et al. (2003) concerning the accelerating impact of the interactions of NO with hydroperoxy radicals via reaction (31). In presence of NO at high concentrations (NO > 100 ppm) the recombination of •OH radicals and NO via reaction (34) was found to be rate governing. This reaction explains the strong inhibiting effect of NO at low temperatures and NO concentrations greater than 100 ppm.

We further performed a coupling of NO with aromatic compounds. The reactions for the coupling of NO$_x$ species with aromatic compounds are listed in Table 13.

Table 13   *Reactions of NO_x compounds and cross-term reactions involved in the oxidation of toluene in presence of NO.*

	Reactions	A	n	E_a	References	No
Reactions of benzene and phenyl	$C_6H_6+NO_2=C_6H_5\bullet+HONO$	7.41×10^{13}	0.0	38200	Chan et al. (2001)	(68)
	$C_6H_6+NO_2=C_6H_5\bullet+HNO_2$	2.5×10^{14}	0.0	42200	Chan et al. (2001)	(69)
	$C_6H_5\bullet+HNO=C_6H_6+NO$	3.78×10^5	2.28	456	Choi et al. (2005)	(70)
	$C_6H_5\bullet+HNO=C_6H_5NO+H\bullet$	3.79×10^9	1.19	95400	Choi et al. (2005)	(71)
	$C_6H_5NO_2=C_6H_5\bullet+NO_2$	1.52×10^{17}	0.0	73717	Xu et al. (2005)	(72)
	$C_6H_5NO=C_6H_5\bullet+NO$	1.52×10^{17}	0.0	55200	Tseng et al. (2004)	(73)
	Reactions of phenyl peroxy radicals					
Reactions of toluene and benzyl radicals	$C_6H_5O_2\bullet+NO=C_6H_5O\bullet+NO_2$	4.7×10^{12}	0.0	-358.0	Estimated[j]	(74)
	benzyl+HNO_2 = toluene+NO_2	8.14×10^4	1.87	4838	Estimated[k]	(75)
	benzyl+HONO = toluene+NO_2	8.14×10^4	1.87	5504	Estimated[l]	(76)
	benzyl+HNO = toluene+NO	1.47×10^{10}	0.76	349.0	Estimated[m]	(77)
	$\bullet C_6H_4CH_3+HNO$ = toluene+NO	3.78×10^5	2.28	456	Estimated[n]	(78)
	benzyl+NO_2= $C_6H_5CH_2O\bullet+$ NO	1.36×10^{12}	0.0	0.0	Estimated[o]	(79)
	$C_6H_5CH_2NO_2$=benzyl+NO_2 *(high pressure)*	1.8×10^{17}	0.0	58500	Estimated[p]	(80)
	(low pressure)	1.3×10^{18}	0.0	42000		
	Reactions of benzylperoxy radicals					
Reactions of ...	$C_6H_5CH_2OO\bullet+$ NO = $NO_2+C_6H_5CH_2O\bullet$	4.70×10^{12}	0.0	-358	Estimated[j]	(81)
	$C_6H_5CH_2O\bullet+NO = C_6H_5CHO+HNO$	7.6×10^{13}	-0.76	0.0	Estimated[q]	(82)
	$C_6H_5CH_2O\bullet+NO_2 = C_6H_5CHO+HONO$	4.0×10^{12}	0.0	2285	Estimated[r]	(83)
	$C_6H_5CH_2O\bullet+HNO = C_6H_5CH_2OH+NO$	3.16×10^{13}	0.0	0.0	Estimated[s]	(84)
	Reactions of benzaldehyde					
	$C_6H_5CHO+NO_2 = C_6H_5CO+HONO$	8.35×10^{-10}	6.68	8300	Estimated[t]	(85)
	$C_6H_5CHO+NO = C_6H_5CO+HNO$	1.02×10^{13}	0.0	40670	Estimated[o]	(86)
Reactions of $HOC_6H_4CH_2\bullet$ radicals	$HOC_6H_4CH_2\bullet+HNO=$ $HOC_6H_4CH_3+NO$	1.47×1011	0.76	349	Estimated[v]	(87)
	$HOC_6H_4CH_2\bullet+NO_2=$ $HOC_6H_4CH_2O\bullet+NO$	1.36×1012	0.0	0.0	Estimated[v]	(88)
	$HOC_6H_4CH_2\bullet+HONO=$ $HOC_6H_4CH_3+NO_2$	8.1×104	1.87	5504	Estimated[w]	(89)
Reactions of $C_6H_5CH_2NO_2$	$C_6H_5CH_2NO_2+\bullet OH=$ $C_6H_5CHO+NO+H_2O$	3.0×10^6	2.0	2000	Estimated[x]	(90)
	$C_6H_5CH_2NO_2+\bullet O\bullet=$ $C_6H_5CHO+NO+\bullet OH$	1.51×10^{13}	0.0	5354	Estimated[x]	(91)
	$C_6H_5CH_2NO_2+H\bullet=$ $C_6H_5CHO+NO+H_2$	4.67×10^{12}	0.0	3732	Estimated[x]	(92)
	$C_6H_5CH_2NO_2+\bullet CH_3= C_6H_5CHO+NO+CH_4$	7.08×10^{11}	0.0	11140	Estimated[x]	(93)
	$C_6H_5CH_2NO_2+\bullet CH_3=$ HCHO+NO+toluene	7.08×10^{11}	0.0	11140	Estimated[x]	(94)
	Reactions of $C_6H_5NO_2$					
	$C_6H_5NO_2=C_6H_5O\bullet+NO$	7.12×10^{13}	0.0	62590	Xu et al. (2005)	(95)
	$C_6H_5NO+NO_2=C_6H_5NO_2+$ NO	9.62×10^{10}	0.0	12928	Park et al. (2002)	(96)

The rate constants are given (k= A T^n exp(-E_a/RT)) in cc, mol, s, cal units.
j : The rate constant was taken as equal to that of reaction (33)
k : The rate constant was taken equal to 1/10th times the value proposed by Dean et al. (1997) for the reactions of •CH$_3$ radicals with HNO$_2$.
l : The rate constant was equal to the value proposed by Dean et al. (1997) for the reaction of HONO with •CH$_3$ radicals.
m : The rate constant was taken as equal to 1/10th times the value proposed by Choi et al (2005) for the reactions of •CH$_3$ radicals with HNO
n : The rate constant was taken as equal to that of reaction (72).
o : The rate constant was taken as equal to that of reaction (49) divided by 10.
p : The rate constant was taken as equal to that of reaction (50).
q : The rate constant was taken as equal to that of reaction (51).
r : The rate constant was taken as the value proposed in GRIMECH 3.0 (Smith et al. 1999) for the reaction CH$_3$O• radicals.
s : The rate constant was taken as equal to that of reaction (58).
t : The rate constant was equal to 10 times the value taken for reaction (53)
u : The rate constant was equal to the value taken for reaction (52).
v : The rate constant was taken equal to that of the similar reaction for benzyl radicals.
w : The rate constant was taken as equal to 1/10th of the value proposed by Dean et al. (1997) for the reactions of •CH$_3$ radicals with HONO.
x : The rate constant was equal to the value proposed in GRIMECH 3.0 (Smith et al. 1999) for the similar reaction of CH$_3$NO$_2$.

For the coupling of NO$_x$ species with aromatic compounds the following reactions were written. There is a great lack of well determined rate constants for reactions between NO$_x$ and aromatic compounds. Thus, most reaction rate constants were estimated by correlations based on molecular structure-reactivity:

- H-abstractions from benzene by NO$_2$ and NO (reactions 68-69) with the rate constants proposed by Chan et al. (2001).

- Reactions of phenyl radicals with HNO (reactions 70-71), NO$_2$ (reaction 72) and NO (reaction 73). Rate constants were taken from recent papers in the literature (Choi et al. 2005, Xu et al. 2005, Tseng et al. 2004).

- Reactions of phenylperoxy (reaction 74) and benzylperoxy (reaction 82) radicals which were derived from analogous reactions of alkylperoxy radicals.

- H-abstractions from toluene by NO$_2$ and NO (reactions 75-78).

- Reactions of resonance stabilized benzyl radicals with NO$_2$ (reactions 79-80).

- Reactions of benzylalkoxy (reactions 82-84) and benzaldehyde, (reactions 85 and 86) with rate constants derived from the corresponding reactions of methoxy radicals and formaldehyde, respectively.

- Reactions HOC$_6$H$_4$CH$_2$ with nitrogen containing species (reactions 87-89) with rate constants derived from the corresponding reactions of benzyl radicals.

- Reactions of C$_6$H$_5$CH$_2$NO$_2$ with small radicals (reactions 90-94) with rate constants derived from the corresponding reactions of CH$_3$NO$_2$.

- Reactions of C$_6$H$_5$NO$_2$ radicals with rate constants taken from Xu et al. (2005) (reaction 95) and Park et al. (2002) (reaction 96).

During our mechanism development, we detected major reaction paths which are determining for the global reactivity of NO sensitized toluene oxidation. lists up the reactions which were found to be rate governing. The corresponding reactions of C$_1$-compounds from which rate constants were derived are shown as well.

Table 14   *Rate governing reactions of NO$_x$ compounds during the oxidation of toluene.*

Reactions	A	n	E_a	References	No
$C_6H_5\overset{\bullet}{C}H_2$ + HONO $\longrightarrow$ $C_6H_5CH_3$ + NO$_2$	8.14x10^4	1.87	5504	Dean et al. (1997)	(76)
$HC\overset{H}{\underset{H}{\overset{\bullet}{C}}}$ + HONO $\longrightarrow$ $HC\overset{H}{\underset{H}{CH}}$ + NO$_2$	8.14x10^4	1.87	5504		(97)
$C_6H_5\overset{\bullet}{C}H$ + NO$_2$ $\longrightarrow$ $C_6H_5\overset{O\bullet}{C}H$ + NO	1.36x10^{12}	0.0	0.0	Srinivasan et al. (2005)	(79)
$HC\overset{H}{\underset{H}{\overset{\bullet}{C}}}$ + NO$_2$ $\longrightarrow$ $HC\overset{H}{\underset{H}{CO\bullet}}$ + NO	1.36x10^{13}	0.0	0.0		(47)
NO$_2$CH$_2$ $\longrightarrow$ $C_6H_5\overset{\bullet}{C}H$ + NO$_2$ (high pressure)	1.8x10^{17}	0.0	58500	*10 x value proposed by Glänzer and Troe (1972).*	(80)
(low pressure)	1.3x10^{18}	0.0	42000		
$HC\overset{H}{\underset{H}{NO_2}}$ $\longrightarrow$ $HC\overset{H}{\underset{H}{\bullet}}$ + NO$_2$ (high pressure)	1.8x10^{17}	0.0	58500		(48)
(low pressure)	1.3x10^{18}	0.0	42000		
$C_6H_5\overset{HO}{C}$ + NO$_2$ $\longrightarrow$ $C_6H_5\overset{O}{C}$ + HONO	8.35x10^{-10}	6.68	8300	Xu et al. (2003)	(85)
$HC\overset{O}{\underset{H}{}}$ + NO$_2$ $\longrightarrow$ $\bullet C\overset{O}{\underset{H}{}}$ + HONO	8.35x10^{-10}	6.68	8300		(51)
$C_6H_5\overset{O\bullet}{\underset{O}{C}}H$ + NO $\longrightarrow$ $C_6H_5\overset{O}{C}H$ + NO$_2$	4.70x10^{12}	0.0	-358	Estimated (see text)	(81)
$R\overset{O}{\underset{O}{}}$ + NO $\longrightarrow$ $R\overset{O}{}$ + NO$_2$	4.70x10^{12}	0.0	-358		(33)

The rate constants are given (k=A T^n exp(-E$_a$/RT)) in cc, mol, s, cal units.

When studying the interactions between NO and toluene, we found a great similarity between the oxidation of methyl radicals in presence of NO and that of benzyl radicals. It was previously described that NO transforms the less reactive methyl peroxy radicals CH$_3$OO$\bullet$ into reactive methoxy radicals CH$_3$O$\bullet$. In case of toluene oxidation, benzyl radicals are produced in large amounts by metatheses. At temperatures below 900 K, benzyl radicals may react with oxygen by addition to oxygen.

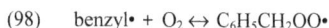

(98)   benzyl$\bullet$ + O$_2$ $\leftrightarrow$ C$_6$H$_5$CH$_2$OO$\bullet$

Compared to linear alkanes the isomerization of the produced peroxybenzyl radicals C$_6$H$_5$CH$_2$OO$\bullet$ is strongly hindered (energies of the surrounding C-H bonds is greater than in the case of alkanes) and the oxidation kinetics of peroxybenzyl radicals C$_6$H$_5$CH$_2$OO$\bullet$ is quite similar to that of methylperoxy radicals $CH_3OO^\bullet$. In presence of NO the rather unreactive $C_6H_5CH_2OO^\bullet$ radical is transformed into a $C_6H_5CH_2O^\bullet$ radical via reaction (81). $C_6H_5CH_2O^\bullet$ radicals easily decompose via reactions:

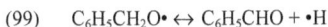

(80) $C_6H_5CH_2NO_2 \leftrightarrow benzyl + NO_2$

(81) $C_6H_5CH_2OO^\bullet + NO \leftrightarrow NO_2 + C_6H_5CH_2O^\bullet$

(99) $C_6H_5CH_2O^\bullet \leftrightarrow C_6H_5CHO + {}^\bullet H$

(100) $C_6H_5CH_2O^\bullet \leftrightarrow C_6H_5{}^\bullet + HCHO$

Decomposition (99) is similar to that of $CH_3O^\bullet$ radicals, which tend to decompose to formaldehyde (reaction (66)). Reaction (81) therefore has a strong accelerating impact on the NO sensitized toluene oxidation.

The described accelerating impact of NO via the transformation of peroxybenzyl radicals is controlled by the net production of benzyl radicals. Reaction (80) is a strong sink of benzyl radicals, especially at low temperatures. Thus, this reaction inhibits global reactivity and controls the accelerating impact of reaction (81). Rate constants of both reactions have to be chosen with care. Rate constants of reactions (80) and (81) were defined identically to those written for the NO interaction of the corresponding methyl compounds and good results were obtained. Reactions (76), (79) and (85) also showed an impact on global reactivity of NO sensitized toluene oxidation.

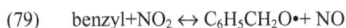

(76) $benzyl + HONO \leftrightarrow toluene + NO_2$

(79) $benzyl + NO_2 \leftrightarrow C_6H_5CH_2O^\bullet + NO$

(85) $C_6H_5CHO + NO_2 \leftrightarrow C_6H_5CO + HONO$

However, their impact on reaction kinetics is rather small compared to the described impact of reactions (80) and (81). In the case of reactions of benzyl with HONO and NO$_2$ or C_6H_5CHO with NO$_2$, rate constants were chosen the same as for the reactions of C_1-species with NO$_x$ compounds.

4.2 Comparison of simulation and previously published experimental data

Validations were performed over a wide range of thermodynamic conditions for different surrogate fuel compositions in various experimental setups. All simulations were run with the commercial software code CHEMKIN IV (Kee et al. 1993). We tested the reaction model against experiments performed in absence and presence of NO.

4.2.1. Model predictivity for hydrocarbon oxidation in absence of NO$_x$

All validations performed in absence of NO are presented in appendix A.2. The model validations against shock tube, rapid compression machines, jet stirred reactor and HCCI-engine experiments generally revealed a good agreement of the model predictions with the experimental data. Low and high temperature oxidation kinetics of the different tested hydrocarbons are well captured over the complete range of tested experimental conditions. Ignition delays and species evolutions are generally correctly reproduced. However, we observed that ignition delays are generally overpredicted. This is true in particular for the case of fuel surrogates including iso-octane. When comparing predicted cool flame ignition delay times of iso-octane to experimental data (Figure A-1b), one observes an over-estimation, despite a correct reproduction of the main ignition delay times. We thus suspect that not all determining decomposition steps during iso-octane oxidation are correctly taken into account in present reaction mechanism.

Nevertheless, the main purpose of our modelling work was the correct reproduction of pollutant formation during postoxidation and for that reason the model is mainly designed for the correct prediction of species evolutions.

4.2.2. Oxidation of PRF and toluene fuels in presence of NO$_x$

The validation of our mechanism against experiments investigating the impact of NO$_x$ on the oxidation of HC is now presented. We remind that given the importance of NO$_x$ on HC oxidation, this is one of the most important aspects of this work. As in the previous section, ignition delays and species profiles were investigated. Experimental data of ignition delays were obtained in HCCI engines and species profiles in a highly diluted JSR. A summary of the tested experimental conditions is shown in Table 15 and Table 16. It is worth noting that no experimental data investigating the impact of the addition of NO on hydrocarbon autoignition in rapid compression machines or shock tube configurations were available.

Table 15   Summary of experimental conditions used for simulations in HCCI-engines.

Author	Exp. setup	Fuel Reference	n-C$_7$ [mol/mol]	FUEL Iso-C$_8$ [mol/mol]	C$_7$H$_8$ [mol/mol]	NO [ppm]	P$_{ini}$ [atm]	T$_{ini}$ [K]	ϕ [-]	Speed [rpm]	CR [-]
	HCCI	Fuel_Dub_1	1.0	0.0	0.0	0-500	1	348	0.3	1500	17
Dubreuil et al. (2007)	HCCI	Fuel_Dub_2	0.75	0.25	0.0	0-500	1	348	0.3	1500	17
	HCCI	Fuel_Dub_3	0.80	0.0	0.2	0-500	1	348	0.3	1500	17
Risberg et al. (2006)	HCCI	Fuel_Ris_1	0.16	0.84	0.0	0-450	1	373	0.25	900	13.6
	HCCI	Fuel_Ris_2	0.35	0.0	0.65	0.450	1	373	0.25	900	13.6

Table 16   Summary of experimental conditions used for auto-ignition simulations in perfectly stirred reactors (PSR).

Author	Exp. setup	Fuel Reference	n-C$_7$ [mol/mol]	FUEL Iso-C$_8$ [mol/mol]	C$_7$H$_8$ [mol/mol]	NO [ppm]	Dilution	P [atm]	T [K]	ϕ [-]
	PSR	Fuel_Morc_1	1.0	0	0.0	0, 50, 500	98 % N$_2$	1	550-1100	1.0
	PSR	Fuel_Morc_1	1.0	0	0.0	0, 50, 500	98 % N$_2$	10	550-1100	1.0
Moréac (2003)	PSR	Fuel_Morc_2	0.0	1.0	0.0	0, 50, 500	98 % N$_2$	1	550-1100	1.0
	PSR	Fuel_Morc_2	0.0	1.0	0.0	0, 50, 500	98 % N$_2$	10	550-1100	1.0
	PSR	Fuel_Morc_3	0.0	0.0	1.0	0, 50, 500	98 % N$_2$	1	550-1100	1.0
	PSR	Fuel_Morc_3	0.0	0.0	1.0	0, 50, 500	98 % N$_2$	10	550-1100	1.0
Dubreuil et al. (2007)	PSR	Fuel_Dub_4	0.15	0.85	0.0	0, 50, 200	98 % N$_2$	10	550-1100	0.2
	PSR	Fuel_Dub_3	0.8	0.0	0.2	0, 50	9 8% N$_2$	10	550-1100	0.2

4.2.2.1. HCCI engines

NO influence on HCCI engine combustion has been studied both by Dubreuil et al. (2007) and Risberg et al. (2006). Simulations were thus performed for these two engine setups. Dubreuil et al. (2007) performed experiments for the oxidation of pure n-heptane, PRF and n-heptane/toluene mixtures with the addition of various concentrations of NO for a test-engine running at 1500 rpm. They observed the evolution of cool flame and main flame ignition delays as a function of the added amount of NO. Ignition delays were obtained by calculating the heat release rate from the recorded pressure history as a function of the Crank

Angle Degree (CAD). It should be mentioned that at the investigated engine speed of 1500 rpm the engine crank turns 1 angle degree in approximately 0.1 ms.

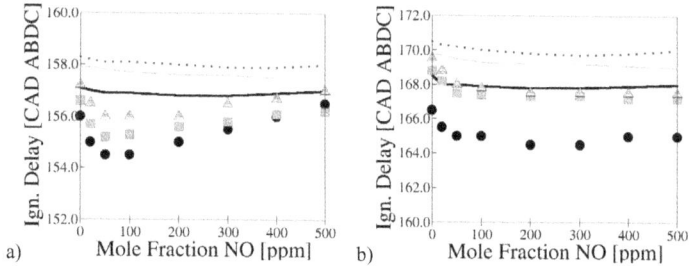

a)   b)

Figure 4-3 *Experimental (symbols) and simulated (lines) ignition delays of (a) cool flame and (b) main flame as function of added NO obtained in an **HCCI-engine** (Dubreuil et al. 2007) for the oxidation of **Fuel_Dub_1** (black circles • and thick full line —), **Fuel_Dub_2** (green squares □ and thin full line -) and **Fuel_Dub_3** (blue shaded triangles △ and dotted line ---).*

Figure 4-3 shows the effect of NO addition on cool flame ignition delays (Figure 4-3a) and main flame ignition delays (Figure 4-3b) for the three fuels referenced in Table 15 (Fuel_Dub_1, Fuel_Dub_2, Fuel_Dub_3). Experiments show that cool flame ignition delays decrease with increasing NO concentrations up to around 100 ppm. For higher concentrations of added NO, the ignition delay increases up to the maximum concentration of 500 ppm without reaching the initial ignition delay obtained in absence of NO. In contrast, the main flame ignition delay decreases strongly for a small addition of NO (up to 50 ppm) while it remains constant for higher NO concentrations. The model shows only slight sensitivity of both, cool flame and main ignition delays to the addition of NO (Figure 4-3a). However, the retarding impact on the ignition delays by blending pure n-heptane (Fuel_Dub_1) with iso-octane (Fuel_Dub_2) and toluene (Fuel_Dub_3) is correctly reproduced (Figure 4-3b). Cool flame and main flame ignition delays are overestimated by less than 10 %.

Risberg et al. (2006) performed experiments on the oxidation on pure PRF and n-heptane/toluene mixtures with the addition of various concentrations of NO for a test engine running at 900 rpm. They traced the heat release evolution as a function of CAD during cool flame and main flame ignition. The heat release rate was obtained from the measured pressure evolution as a function of the added amount of NO.

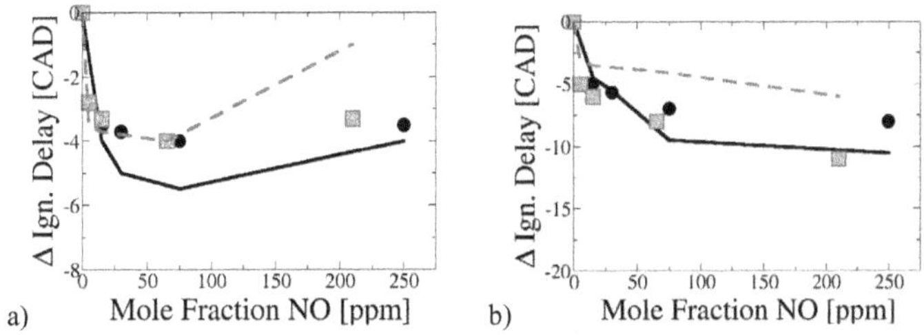

*Figure 4-4 Experimental (symbols) and simulated (lines) sensitivity to NO addition of (a) cool flame and (b) main flame in [CAD] obtained in an **HCCI-engine** (Risberg et al. 2006) for the oxidation of **Fuel_Ris_1** (black circles • and thick full line —), **Fuel_Ris_2** (green squares □ and dotted lines ---).*

Figure 4-4 shows the effect of NO addition on cool flame ignition delays (Figure 4-4a) and main flame ignition delays (Figure 4-4b) for the two fuels referenced in Table 15 (Fuel_Ris_1, Fuel_Ris_2). In both cases, the ignition delay time has been determined from the heat release profile by taking the CAD value at half of the reached maximum value of the heat release. In the case of the PRF-mixture, as well as in the case of the n-heptane/toluene mixture, the experiments show a decrease of the cool flame ignition delays when the NO concentrations are increased up to 75 ppm. For this NO concentration, the cool flame ignition delays reach a minimum and are retarded for higher NO contents. This is not the case for the main flame ignition delays which decrease over the complete range of tested NO concentrations. To model these experimental results, we adjusted the initial temperature such that ignition delays in absence of NO were correctly reproduced. We then simulated the impact of NO addition by keeping initial thermodynamic conditions constant. This procedure corresponds to a validation of the model sensitivity against NO addition. Our model captures well the effect of the NO addition on the cool flame and main flame ignition delays for both fuels. Nevertheless, the model predicts a higher sensitivity to NO for PRF-mixtures than for n-heptane/toluene mixtures which is in contrast with experimental results.

4.2.2.2. Jet Stirred Reactors (JSR)

Moréac et al. (2003) also studied the impact of NO on the oxidation of n-heptane, iso-octane and toluene at different pressures and at temperatures from 500 to 1100 K. Figure 4-5 to Figure 4-8 show the model validations against these experimental data.

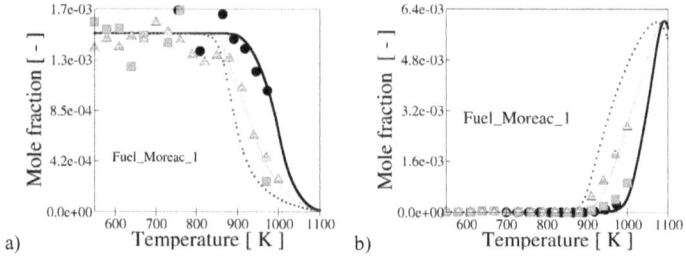

*Figure 4-5   Experimental (symbols) and simulated (lines) concentration profiles of (a) n-heptane mole fractions and (b) CO mole fractions obtained in a **JSR** (Moréac et al. 2003) for the stoichiometric oxidation of 1500 ppm of **n-heptane** (**Fuel_Moreac_1**) at a pressure of **1 atm** for an addition of 0 ppm NO (black circles ● and thick full line ━), 50 ppm NO (green squares ▪ and thin full line ·) and 500 ppm NO (blue shaded triangles △ and dotted lines ---).*

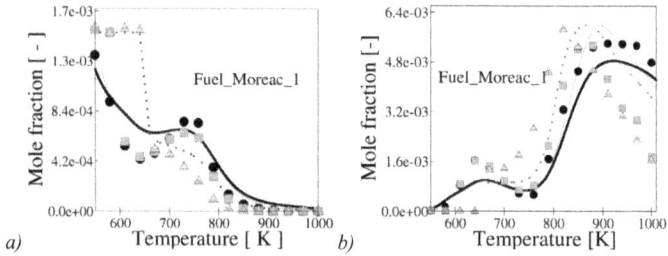

*Figure 4-6   Experimental (symbols) and simulated (lines) concentration profiles of (a) n-heptane mole fractions and (b) CO mole fractions obtained in a **JSR** (Moréac et al. 2003) for the stoichiometric oxidation of 1500 ppm of **n-heptane** (**Fuel_Moreac_1**) at a pressure of **10 atm** for an addition of 0 ppm NO (black circles ● and thick full line ━), 50 ppm NO (green squares ▪ and thin full line ·) and 500 ppm NO (blue shaded triangles △ and dotted line ---).*

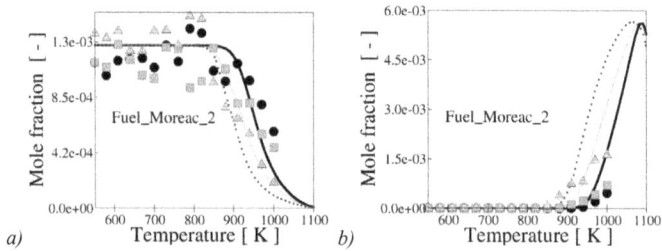

*Figure 4-7   Experimental (symbols) and simulated (lines) concentration profiles of (a) iso-octane mole fractions and (b) CO mole fractions obtained in a **JSR** (Moréac et al. 2003) for the stoichiometric oxidation of 1250 ppm of **iso-octane** (**Fuel_Moreac_2**) at a pressure of **1 atm** for an addition of 0 ppm NO (black circles ● and thick full line ━), 50 ppm NO (green squares ▪ and thin full line ·) and 500 ppm NO (blue shaded triangles △ and dotted line ---).*

Figure 4-5 (n-heptane) and Figure 4-7 (iso-octane) show that, at atmospheric pressure, the experimentally observed promoting impact of NO on both the conversion of reacting alkane (Figure 4-5a and Figure 4-7a) and the CO concentration profiles (Figure 4-5b and Figure 4-7b) are captured well by the model. Experimental results presented in Figure 4-6 for the oxidation of n-heptane at a pressure of 10 atm reveal a complex impact of NO addition. The addition of NO in small concentrations (50 ppm) shows a slight retarding effect on the minimum temperature at which n-heptane starts to react. In contrast, the addition of larger amounts of NO (500 ppm) causes a strong inhibition and the minimum temperature above which n-heptane reaction is detected shifts up to 650 K. One also observes that the NTC-effect between 650 and 750 K is reduced by small amounts of added NO (50 ppm) and completely disappears, when NO is present in larger concentrations (500 ppm). Above 750 K and a pressure of 10 atm the addition of NO generally promotes the oxidation of n-heptane stronger than at atmospheric pressure. Neither the inhibition of the n-heptane oxidation at low temperatures, nor the acceleration by NO at higher temperatures is a linear function of the amount of added NO. The comparison of the simulation results with the experimental data shows that the model predicts the inhibiting impact of NO at low temperature, as well as its promoting effect at higher temperature. In addition the reduction of the NTC effect for small concentrations of NO (50 ppm) and the sudden decrease of the n-heptane concentrations for the addition of 500 ppm of NO at 650 K are well captured.

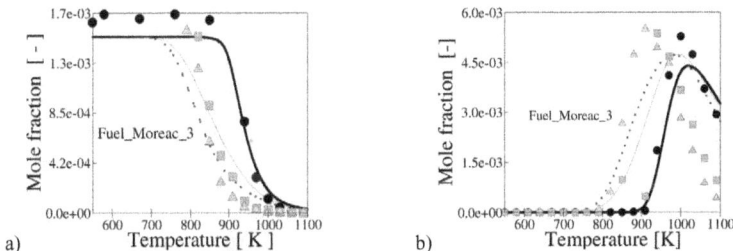

a)

b)

*Figure 4-8 Experimental (symbols) and simulated (lines) concentration profiles of (a) toluene mole fractions and (b) CO mole fractions obtained in a JSR (Moréac et al. 2003) for the stoichiometric oxidation of 1250 ppm of **toluene (Fuel_Moreac_3)** at a pressure of **10 atm** for an addition of 0 ppm NO (black circles ● and thick full line ▬), 50 ppm NO (green squares □ and thin full line -) and 500 ppm NO (blue shaded triangles △ and dotted line ---).*

The experimental results for the oxidation of toluene displayed in Figure 4-8 show that the addition of small amounts of NO (50 ppm) leads to a shift of the toluene and CO concentration profiles towards lower temperatures by 30 to 50 K. An addition of 500 ppm of NO results in further shifts of the toluene and CO concentration profiles of 40 to 60 K and the acceleration of the toluene oxidation by NO is non-linear with respect to the amount of NO added. The kinetic model captures well the shift in toluene and CO concentration profiles, but the reactivity of toluene in presence of NO is underestimated. Dubreuil et al. (2007) performed as well experiments in a JSR for different fuel mixtures and several contents of NO.

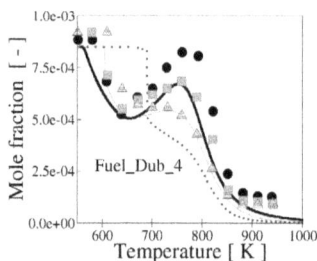

*Figure 4-9 Experimental (symbols) and simulated (lines) concentration profiles of iso-octane mole fractions obtained in a **JSR** (Dubreuil et al. 2007) for the lean oxidation (ϕ=0.2) of **Fuel_Dub_4** at a pressure of **10 atm** for an addition of 0 ppm NO (black circles ● and thick full line ▬), 50 ppm NO (green squares □ and thin full line -) and 200 ppm NO (grey shaded triangles △ and dotted line ---)*

Variations of recorded NO and NO$_2$ concentrations are indicated between ±5-20 ppm. As shown in Figure 4-9, the impact of NO on the PRF oxidation observed experimentally is similar to that obtained by Moréac et al. (2003) for the oxidation of pure n-heptane at 10 atm (compare Figure 4-6a). The addition of NO inhibits the oxidation of PRF mixtures below 700 K, while it promotes it at higher temperatures. The NTC-effect around 750 K is reduced by the addition of small amounts of NO (50 ppm) and disappears completely when NO is added in higher concentration (200 ppm). The comparison of experimental results to simulations shows that the described effects of NO addition are well captured by the model. However, for high concentrations of NO (200 ppm) the model overestimates the minimum temperature above which the reactivity of iso-octane is detected.

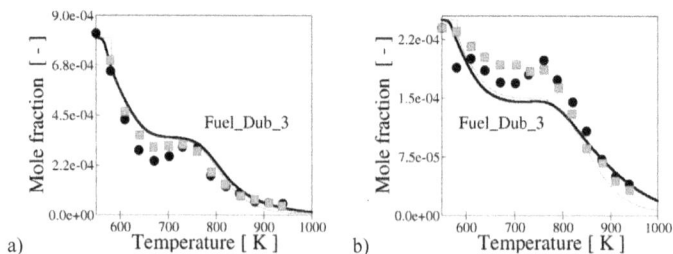

a)   b)

*Figure 4-10   Experimental (symbols) and simulated (lines) concentration profiles of (a) n-heptane mole fractions and (b) toluene mole fractions obtained in a **JSR** (Dubreuil et al. 2007) for the lean oxidation (ϕ=0.2) of **Fuel_Dub_3** at a pressure of **10 atm** for an addition of 0 ppm NO (black circles ● and thick full line ▬) and 50 ppm NO (green squares □ and thin full line -).*

Experimental results presented in Figure 4-10 for the oxidation of n-heptane/toluene show that the addition of NO causes an inhibition of the n-heptane/toluene oxidation at temperatures below 700 K, while at higher temperatures a promoting effect is observed. These effects are reproduced well by the model.

4.2.3. Model predictivity for hydrocarbon oxidation in presence of NO$_x$

The written sub-mechanism of NO$_x$ was validated against jet stirred reactor and HCCI-engine experiments. Satisfactory agreement of the model predictions compared to the experimental data was obtained over a large range of experimental conditions. The written reactions between NO$_x$ compounds and hydrocarbons lead to a good prediction of the NO interactions on hydrocarbon oxidation. The rate governing reactions of NO$_x$ compounds with hydrocarbons were described in the previous section. The impact of NO on hydrocarbon oxidation at different temperatures (low temperature: inhibiting / high temperature: accelerating) is captured correctly for any tested experimental setup. The kinetic sensitivity against different concentrations of NO is satisfactory reproduced as well. The NO-sensitivity was tested under lean and stoichiometric conditions. However, a validation under rich conditions is missing, as no such experimental data was available.

4.3   Sensitivity and flux analysis for NO$_x$ compounds containing reactions

Sensitivity and reaction rate analyses were performed with the previously described mechanism for JSR-simulations of neat n-heptane and toluene oxidation for various temperatures (665 K, 750 K, 950 K), different pressures (1 atm, 10 atm) and two amounts of added NO (50 ppm, 500 ppm). The simulation conditions correspond to the experimental conditions chosen by Moréac et al. (2003) for the previously shown experiments on the oxidation of pure hydrocarbons in a JSR (Table 11). The relative mol reaction rates indicated in Figure 4-11 and Figure 4-14 for a reaction x represent the mol fluxes via reaction x normalized by the rate of production of heptyl by benzyl radicals, respectively. Sensitivity coefficients σ were obtained by the following formula:

$$[5] \quad \sigma_x = \frac{M_{reac\,tan\,ts}\left(k_x \cdot 10\right) - M_{reac\,tan\,ts}\left(k_x/10\right)}{9.9 \cdot M_{reac\,tan\,ts}\left(k_x\right)}$$

where $M_{reactant}$ $(k_x \times 10)$ is the reactant mole fraction obtained for a simulation run with the rate constant of reaction x multiplied by a factor 10, $M_{reactant}$ $(k_x/10)$ the reactant mole fraction obtained for a simulation run with the rate constant of reaction x divided by a factor 10 and $M_{reactant}$ (k_x) the reactant mole fraction simulated by the initial mechanism. A positive sensitivity coefficient σ_x indicates an inhibitive effect and a negative coefficient shows an accelerating impact of reaction x on the global reactivity.

4.3.1. Oxidation of n-heptane

Figure 4-11 to Figure 4-13 show the reaction rate and sensitivity analyses performed for the oxidation of n-heptane.

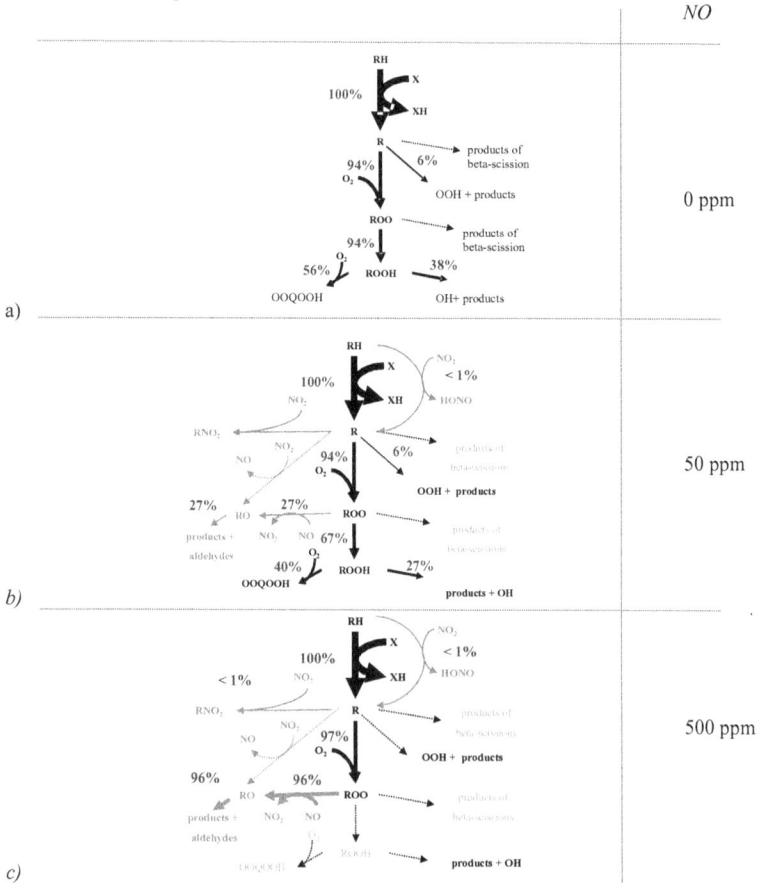

Figure 4-11   Flux analysis for the stoichiometric oxidation of 1500 ppm of **n-heptane** in a PSR at a pressure of **10 atm**, a temperature of **665 K**, and a residence time of **1 s** with addition of (a) 0 ppm of NO, (b) 50 ppm of NO and (c) 500 ppm of NO.

Sensitivity Coefficient

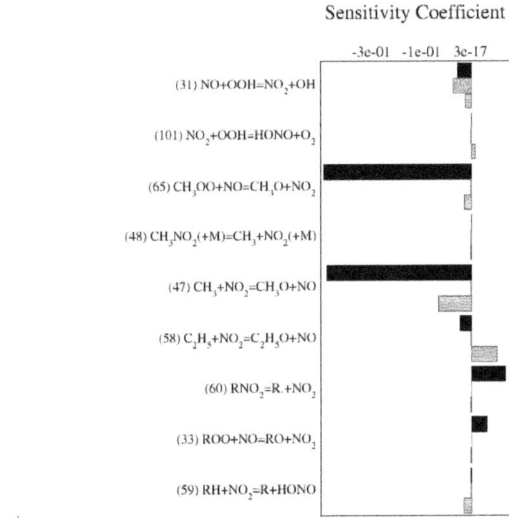

a)

Sensitivity Coefficient

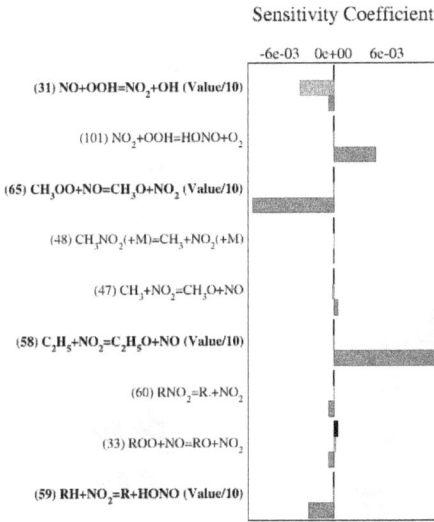

b)

*Figure 4-12   Sensitivity analysis for the stoichiometric oxidation of 1500 ppm of **n-heptane** in a PSR at a pressure of **1 atm** at a temperature of 665 K (black bars), 750 K (green bars) and 900 K (blue bars) for (a), an initial concentration of NO of 50 ppm and (b) an initial concentration of NO of 500 ppm.*

Sensitivity Coefficient

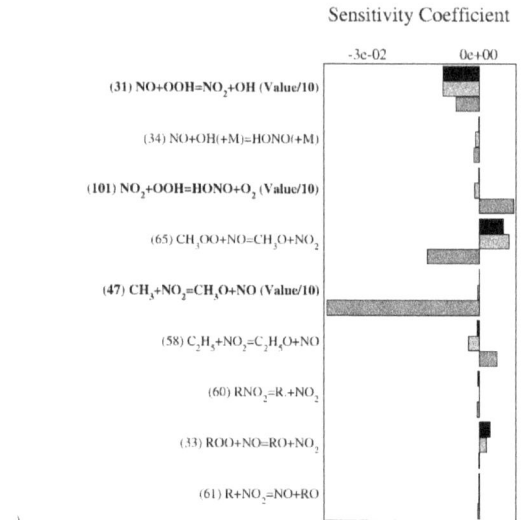

a)

Sensitivity Coefficient

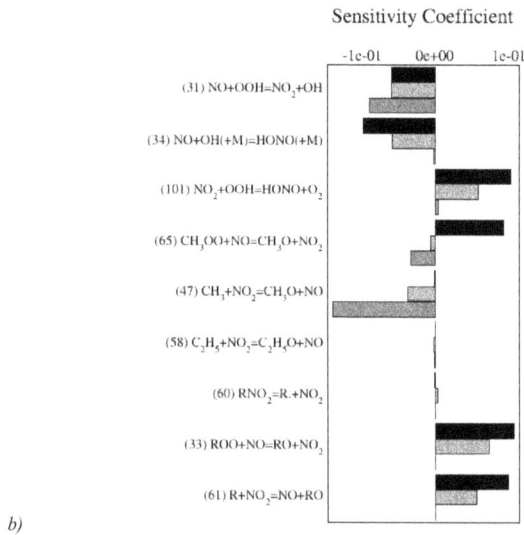

b)

*Figure 4-13 Sensitivity analysis for the stoichiometric oxidation of 1500 ppm of **n-heptane** in a PSR at a pressure of **10 atm** at a temperature of 665 K (black bars), 750 K (green bars) and 900 K (blue bars) for (a), an initial concentration of NO of 50 ppm and (b) an initial concentration of NO of 500 ppm.*

Figure 4-11a shows the classical scheme for the oxidation of an alkane at low temperatures (665 K) in absence of NO$_x$. The main reaction channel is the formation of alkyl (R•) radicals from the initial reactant followed by an addition of oxygen molecules and an isomerization of the obtained peroxy radicals (ROO•) to give hydroperoxyalkyl (•QOOH) radicals. These radicals can decompose into stable species, such as cyclic ethers or ketones, involving the expulsion of •OH radicals. They can also react by the addition with another oxygen molecule producing hydroperoxyalkylperoxy (•OOQOOH) radicals. The isomerization and decomposition of •OOQOOH radicals lead to the formation of hydroperoxide molecules. The decomposition of these hydroperoxide molecules involves a multiplication of the radical production, which in a chain reaction induces an exponential acceleration of the reaction rate. At temperatures around 750 to 800 K, the reversibility of the addition of alkyl (R•) radicals to oxygen molecules becomes more important. The oxidation of these radicals leading to the formation of alkenes and the very unreactive •OOH radicals is then favoured. This reduces the overall reactivity and is the main reason for the appearance of the NTC regime.

The presence of NO$_x$ considerably changes the reaction scheme of the oxidation of alkanes. This is illustrated in Figure 4-11b and Figure 4-11c. The reaction of peroxy radicals with NO gives alkoxy (RO•) radicals and NO$_2$ (reaction (33) in Table 11). The resulting RO• radicals are decomposed to aldehydes and smaller alkyl radicals (reaction 64). This reaction channel competes with the second addition of oxygen molecules and thus reduces the rate of formation of hydroperoxide species disadvantaging the branching steps which are induced by the decomposition of •OOQOOH radicals. With an increased amount of added NO, a considerable rise of the RO• production is observed. For an addition of 500 ppm of NO at 10 atm and 665 K, 96 % of peroxy radicals are converted to alkoxy radicals RO• (Figure 4-11c). This explains the inhibiting impact of NO$_x$ on hydrocarbon oxidation between 650 and 750 K. The sensitivity analyses displayed in Figure 4-12 and Figure 4-13 reveals that reaction (33) has an inhibiting impact on the n-heptane oxidation under any investigated conditions. This inhibition is particularly important at 650 K and a pressure of 10 atm. It can be noted that the reactions of hydroperoxyalkylperoxy radicals (HOOQOO•) and NO are always of negligible importance and could have been omitted. At low temperatures, a small proportion of alkyl radicals may react with NO$_2$ yielding RNO$_2$ molecules (reaction 60) or alkoxy radicals and NO (reaction 61). The production of RNO$_2$ molecules is mainly important at atmospheric pressure and temperatures below 750 K, where it strongly inhibits the global reactivity, thus compensating the accelerating impact of the reactions of CH$_3$OO• (reaction 65) and •CH$_3$ (reaction 47) with NO$_x$ (Figure 4-12a). At higher pressures (10 atm) the production of alkoxy radicals by reaction (65) has also a noticeable effect. This reaction competing with the addition to oxygen molecules has an inhibiting effect, which is particularly important at 10 atm and 750 K (Figure 4-13b). The reaction between alkanes and NO$_2$, leading to nitrous acid (HONO) (reaction 14) has a slight impact at 1 atm and 665 K. The H-abstractions from aldehydes by NO$_2$ (reaction 63) and the reaction of resonance stabilized (Y•) radicals with NO$_2$ (reaction 64) producing acrolein, NO and alkyl radicals show only a limited effect on the overall reactivity. Above the NTC zone (900 K), reactions of alkyl radicals with oxygen molecules become less important and the influence of reactions (41, 59-64) is almost negligible. Sensitivity analyses further show that the reactions of NO$_x$ with C_1-C_2 species are of great importance whatever the temperatures range. Under almost any condition of hydrocarbon oxidation, reactions (47), (31), (65), and (101) show large sensitivity coefficients:

$$(49) \quad \bullet CH_3 + NO_2 \quad \leftrightarrow \quad CH_3O \bullet + NO$$

$$(31) \quad NO + \bullet OOH \quad \leftrightarrow \quad NO_2 + \bullet OH$$

$$(67) \quad CH_3OO \bullet + NO \quad \leftrightarrow \quad CH_3O \bullet + NO_2$$

$$(101) \quad NO_2 + \bullet OOH \quad \leftrightarrow \quad HONO + O_2$$

The promoting reactions (31) and (47) are particularly important at atmospheric pressure and at low temperatures, where the mechanism is very sensitive to the addition of NO. At low temperatures, the general fate of •OOH radicals is their recombination to form H$_2$O$_2$. Methyl radicals (•CH$_3$) lead to the production of rather stable CH$_3$OO• radicals via the addition of O$_2$. Reaction (31) transforms the unreactive •OOH into the very reactive •OH radicals, while reaction (47) produces reactive CH$_3$O• radicals from •CH$_3$ radicals, thus increasing the overall reactivity. Reaction (101) is a termination step competing with reaction (31) and showing generally an inhibiting effect. The impact of the reaction of CH$_3$OO• radicals with NO (reaction (65)) is more complex. At 10 bar it inhibits the global reactivity at 665 K, but accelerates it at temperatures above 900 K (Figure 4-13). The inhibiting impact of reaction (65) at low temperature might be caused by its competition with the disproportionation of CH$_3$OO• with •OOH radicals yielding the branching agent, CH$_3$OOH, and oxygen molecule. At higher temperatures, this disproportionation becomes less important and thus reaction (65) has an accelerating impact. It should be noted that the contribution of the recombination of •CH$_3$ radicals and NO$_2$ via reaction (47) is negligible compared to the impact of larger alkyl radicals reacting via the analogous reaction type (61). Below 650 K, at a pressure of 10 atm and for a high concentration of NO (500 ppm), the alkane oxidation is governed by reaction (34).

(34)   NO+•OH+(M)   ↔   HONO+(M)

(61)   R•+NO$_2$ →   RO•+NO

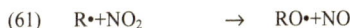

This explains the strong inhibition of n-heptane oxidation in presence of NO at temperatures below 650 K shown by Figure 4-6a. At these temperatures, reactive •OH radicals are consumed via reaction (34) producing nitrous acid (HONO) and representing thus an important sink for •OH radicals. At temperatures above 650 K, the dissociation of HONO exceeds its production and the sense of reaction (34) is reversed. The reverse of sense of reaction (34) causes a sudden acceleration of the overall reactivity and provokes the sudden decrease of the reactant concentration profile shown in Figure 4-6a. However, under same conditions, but at atmospheric pressure, no reactivity is observed, neither in presence nor in absence of NO (Figure 4-5a). One other important reaction is reaction (58) describing the reaction of ethyl (•C$_2$H$_5$) radicals with NO$_2$. At low temperature, in absence of NO, the main reactions of ethyl radicals are with oxygen molecules to give the rather unreactive C$_2$H$_5$OO• radical. Reaction (58) provides an alternative channel towards the more reactive ethoxy (C$_2$H$_5$O•) radical, explaining thus their promoting effect. Above 800 K, reaction (58) becomes inhibiting: at these temperatures, •C$_2$H$_5$ radicals react mainly with O$_2$ to give ethylene and •OOH radicals, which again are transformed to •OH radicals via reaction (31). The competing reaction (58) thus has an inhibiting impact, as it produces stable ethoxy radicals instead of the easily converted •OOH radicals.

4.3.2. Oxidation of toluene

Figure 4-14 and Figure 4-15 show, respectively, the reaction rates and sensitivity analyses performed for the oxidation of toluene at 900 K in the case of the reaction of aromatic compounds. Figure 4-15 shows the sensitivity coefficients of governing reactions containing aromatic compounds and NO$_x$. Bounaceur et al. (2005) showed that at around 900 K, in absence of NO, toluene is mainly consumed to give resonance stabilized benzyl radicals. Thus, all relative mol fluxes indicated in Figure 4-14 are normalized to the rate of production of benzyl radicals. Benzyl radicals mainly react by combination with themselves to give bibenzyl or with •OOH radicals producing benzylhydroperoxide molecules and rapidly yielding alkoxybenzyl radicals which decompose to produce benzaldehyde and H• atoms or formaldehyde and phenyl radicals. A minor channel involves the reaction of benzyl radicals with oxygen molecules to give peroxybenzyl radicals, which isomerize and decompose to give benzaldehyde and H• atoms.

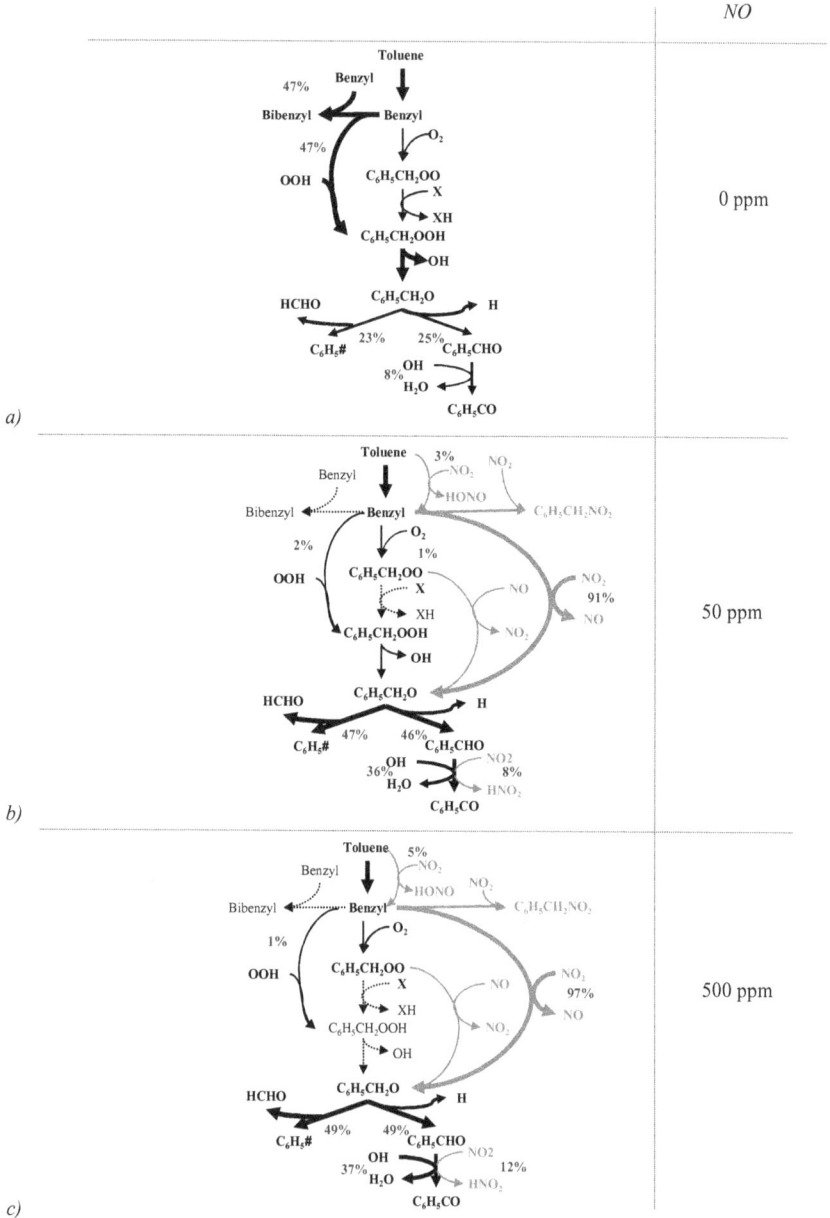

*Figure 4-14   Flux analysis for the stoichiometric oxidation of 1500 ppm of **toluene** in a PSR at a pressure of **10 atm**, a temperature of **900 K** and a residence time of 1 s with addition of (a) 0 ppm of NO, (b) 50 ppm of NO and (c) 500 ppm of NO.*

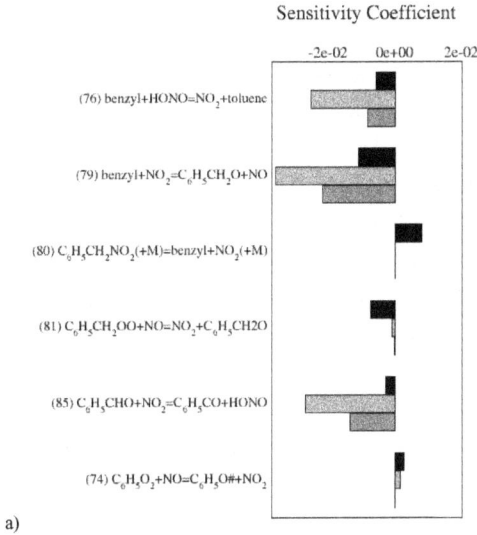

a)

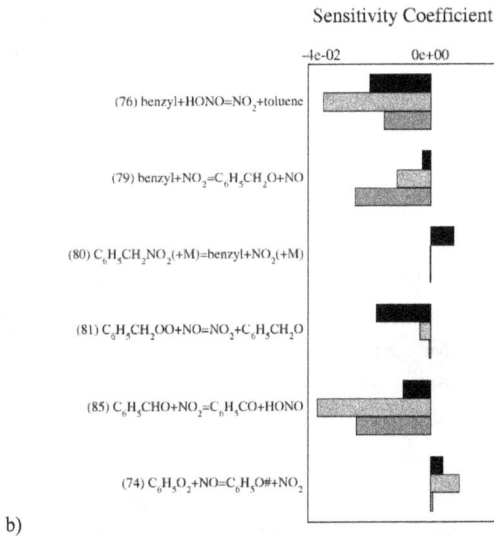

b)

*Figure 4-15   Sensitivity analysis for the stoichiometric oxidation of 1500 ppm of **toluene** in a PSR at a pressure of **10 atm** at a temperature of 750 K (black bars), 900 K (green bars) and 1000 K (blue bars) for (a), an initial concentration of NO of 50 ppm and (b) an initial concentration of NO of 500 ppm.*

In the presence of nitrogen containing species, 5 % of toluene is consumed by H-abstractions with NO$_2$ producing HONO and benzyl radicals (reaction 76). Due to the rapid dissociation of HONO (reaction (34)) giving •OH radicals and NO, this reaction has an important promoting effect in the studied temperature range, even if it leads to the formation of resonance stabilized benzyl radicals. Benzyl radicals can react with NO$_2$ to give directly alkoxybenzyl radicals and NO (reaction 79) and this reaction shows an important promoting effect. The combination of benzyl radicals with NO$_2$ (reaction 80) competes with reaction (79) and has an inhibiting effect. Peroxybenzyl radicals, which tend not to easily isomerize, can react with NO and produce alkoxybenzyl radicals (reaction 81) promoting the production of alkoxybenzyl. Nevertheless, in the presence of NO$_x$ the importance of reaction (79) is determinant: according to Figure 4-14b and Figure 4-14c more than 90 % of benzyl radicals are consumed via this channel. In comparison, the effect of reaction (81) is rather limited, as less than 1 % of peroxybenzyl radicals are consumed via this reaction channel. One observes that the H-abstractions from benzaldehyde by NO$_2$ (reaction 85) has a promoting effect, while the reaction of peroxyphenyl with NO producing resonance stabilized phenoxy radicals and NO$_2$ (reaction 74) has an inhibiting impact.

4.4   Evaluation of the developed detailed kinetic mechanism

The kinetic model for the oxidation of a PRF/toluene blends presented here has been successfully validated against different experimental applications over a wide range of thermochemical conditions. Ignition delays obtained in rapid compression machine, shock tube and HCCI experiments and concentration profiles measured in JSR experiments have been modelled. A model describing the impact of NO$_x$ on hydrocarbon oxidation has been developed and coupled with the PRF/toluene mechanism. This model was validated against HCCI and PSR experiments for neat fuel and various fuel blends. Validations show that the model accurately captures the complex impact of NO on hydrocarbon oxidation. The impact at different pressures and varying temperatures for various concentrations of NO is well retrieved for all tested fuels. The sub-model containing nitrogen species was analysed by sensitivity and flux analyses and the important reaction channels have been identified permitting a deeper understanding of the impact of NO$_x$ on hydrocarbon oxidation. The good results of model validations for different experimental setups over a wide range of thermochemical conditions should allow the use of the proposed mechanism for IC-engine applications, as well as for postoxidation applications governing gas flows in the exhaust line.

Nevertheless, the actual reaction model may be improved in future works. For the correct prediction of ignition delay times at high pressures ($P > 40$ bar) the user has to change rate coefficients of the H_2O_2-dissociation (compare section 4.1.1.2). This might be due to uncertainties of collision efficiencies of collision partner in 3rd-body reactions (see studies presented in the following section 5). More detailed studies on the kinetics of C_0-C_1-species are necessary for an optimized prediction of low temperature hydrocarbon oxidation kinetics.

Concerning the oxidation of iso-octane, we had to divide the rate coefficients of the iso-octyl oxidation reactions by a factor of 3 in order to obtain a convenient prediction of iso-octane ignition delay times. This is not coherent with the EXGAS approach generating rate constants automatically by an additivity method. Compared to n-heptane, iso-octane shows quite particular oxidation characteristics. When ignited in shock tubes or rapid compression machines iso-octane ignition delay times show a well expressed NTC-behaviour at temperatures between 700 and 850 K. However, the experimental results performed in jet-stirred reactors show not any reactivity of iso-octane at temperatures below 850 K. This is not the case for n-heptane, which shows a well expressed NTC-zone in shock tubes or rapid compression machines as well as in jet stirred reactors. Nevertheless iso-octane has a tendency to produce cool flames (compare Figure A-1b) and the difference between cool and main ignition delay times is greater for iso-octane than for n-heptane (compare (Figure A-1a).

Chapitre 5

Numerical studies of reaction kinetics under postoxidation conditions

The new developed model which was described in the previous chapter has been used to further study chemical kinetics under postoxidation conditions. Our literature review has revealed a great lack of information on the impact of the dilutant composition on the oxidation of large hydrocarbons. To our knowledge, the impact of the major components of burned gases, such as CO, CO_2 and H_2O on fuel oxidation has not been tested under high dilution conditions and at low temperatures between 600 and 1000 K. This motivated us to perform such a study with the developed kinetic model. It is reminded that it has been extensively validated in a wide range of thermodynamic conditions. Our results were compared to those obtained by the model of Naik et al. (2005b).

We had observed as well that our model tends to overestimate the production of ethylene. We have therefore improved the reactions producing ethylene which are defined in the secondary mechanism of the EXGAS-software (Warth et al. 1998). This has been done by studying the oxidation of n-heptane/toluene blends. We have developed a new mechanism for the oxidation of n-heptane/toluene blends which is based on that presented in chapter 4 for the oxidation of PRF/toluene mixtures. That n-heptane/toluene mechanism was validated against experimental results obtained by Piperel in a Jet Stirred Reactor (JSR) and described in the paper of Anderlohr et al. (2009a). In that context, we have also studied numerically the impact of CO, HCHO and C_2H_6 on hydrocarbon oxidation.

The work presented in this chapter is the object of two papers. The first part corresponds to a paper accepted by the Combustion, Science and Technology journal:

(Anderlohr JM., Pires da Cruz A, Bounaceur R, Battin-Leclerc F, (2010). Thermal and kinetic Impact of CO, CO₂ and H₂O on the Postoxidation of IC-Engine Exhaust Gases.
Combst. Sci. Technol, 182, 30-59)

and the second has been published in the Proceedings of the Combustion Institute:

(Anderlohr JM., Piperel A, Pires da Cruz A, Bounaceur R, Battin-Leclerc F, Dagaut P,
Montagne X, (2009). Influence of EGR Compounds on the oxidation of an HCCI-Diesel surrogate.
Proc. Combust. Inst., 32, 2851-2859.)

A third part of this chapter presents a study of the thermodynamic conditions under which auto-ignition may occur in an IC-engine exhaust line.

5.1   Study on the impact of residual gases on hydrocarbon oxidation

In this first section, the thermal and kinetic impact of the residual gases on hydrocarbon oxidation chemistry is numerically investigated. The case of pure dilution by N_2 was tested against a dilutant composed of CO, CO_2 and H_2O in proportions corresponding to IC-engine postoxidation conditions (at the end of the expansion stroke and throughout exhaust). The impact of each residual species was tested individually, as well as in combination with others.

Attention was given to the thermal impact, kinetic impact and third body effects of each residual gas. In the cases of CO_2 and H_2O, a negative thermal C_p-effect in competition to an accelerating kinetic impact due to 3rd-Body reactions was observed. In this section, the impact of CO, CO_2 and H_2O on the autoignition delays of hydrocarbons is investigated at thermodynamic conditions typical of fuel postoxidation in engines: a high dilution ratio (97%) and low to medium temperatures (650-1100 K). This is the temperature range where the oxidation chemistry is the most complex. The 'ignition delay' is here defined as the residence time for which 50% of the maximum oxidation heat is released.

5.1.1. Simulation conditions and results

According to the conditions found in engine exhaust manifolds, this study was mainly performed at *constant atmospheric pressure*. The reactor type chosen was a closed homogeneous adiabatic reactor. All simulations were made using the commercial Chemkin 4.1 (Kee et al. 1993) software package. We used in this study the reaction mechanism which was previously described in chapter 4. The chosen fuel surrogate consists of a mixture of 13.7 mol% n-heptane, 42.9 mol% iso-octane and 43.4 mol% toluene and corresponds to a gasoline surrogate of Research Octane Number (RON) of 95. Simulations were run for a large range of equivalence ratios ϕ and initial temperatures T. Table 17 shows the thermodynamic conditions covered by this study.

Table 17   Thermochemical conditions during postoxidation: a) typical composition of a fuel surrogate, b) thermodynamic conditions.

a)	Fuel composition			b)	Thermodynamic conditions		
	n-heptane	*iso-octane*	*toluene*		*P*	*T*	ϕ
	$[mol/mol]$	$[mol/mol]$	$[mol/mol]$		[bar]	[K]	[-]
	13.7	42.6	43.7		1	650-1100	0.1-7

Those simulations lead to the establishment of an ignition delay cartography in the T-ϕ space shown on Figure 5-1. It is observed that ignition delays as defined in this section are very long when compared to internal combustion engines characteristic time scales (in the order of tenths of a millisecond).

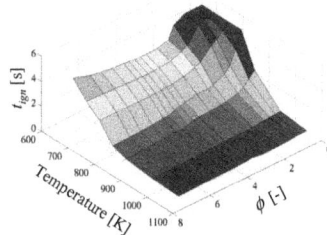

Figure 5-1   *Ignition delay times as a function of initial temperature T and equivalence ratio ϕ for the oxidation of the n-heptane/iso-octane/toluene mixture, diluted by 97 % of N_2 (Reference-case).*

For all simulations, the dilution ratio was held constant at 97 mol%. Under such regime, ignition delays range in an order of magnitude of seconds. In order to evaluate the impact of CO, CO_2 and H_2O on fuel oxidation, simulation runs were first performed with pure N_2 as the residual species (*Reference*-case). Then, the case of pure N_2-dilution was compared to a more engine realistic residual composition (*Multiple-dilutant*-case) composed of CO, CO_2 and H_2O. The composition of the *Multiple-dilutant*-case was chosen according to exhaust gas measurements performed at the IFP test-benches on a gasoline engine at stoichiometric conditions. In order to evaluate individually the impact of each major compound present in the dilutant on hydrocarbon oxidation, we have replaced the N_2 in the *Reference-Case* by 100% of each of the investigated components. Then, we have analysed the impact of each tested compound separately at concentrations found during postoxidation. This was made by keeping constant the concentrations of each investigated species, such as defined in the *Multiple-Dilutant*-case, and replacing all others by N_2. Table 18 summarizes the mixture compositions for all simulations described in this section.

Table 18   *Diluting compositions for the study of the influence of various residual species on hydrocarbon oxidation.*

Case	N_2 $[mol/mol]$	CO $[mol/mol]$	CO_2 $[mol/mol]$	H_2O $[mol/mol]$	Dilution $[mol/mol]$
Reference	0.97	0.0	0.0	0.0	0.97
Multiple Dilutant	0.685	0.005	0.14	0.14	0.97
Single Dilutant	0.0	0.97	0.0	0.0	0.97
	0.0	0.0	0.97	0.0	0.97
	0.0	0.0	0.0	0.97	0.97
Binary Dilutant	0.965	0.005	0.0	0.0	0.97
	0.83	0.0	0.14	0.0	0.97
	0.83	0.0	0.0	0.14	0.97

Reference-Case and Multiple Dilutant-case

We first compared the Reference-case (Table 18: Dilution by pure N_2) against the *Multiple-Dilutant*-case (Table 18: Dilution by a N_2, CO, CO_2, H_2O mixture). Figure 5-2 compares the temperature evolutions during fuel oxidation at three different initial temperatures ($T = 650$, 750 and 950 K) and equivalence ratios ($\phi = 0.4$, 1.0 and 4.0). The temperatures shown are plotted against the normalized time τ, which was calculated as:

$$[6] \quad \tau = \frac{t}{t_{AI(N2)}}$$

Where $t_{AI(N2)}$ is the ignition delay of the N_2 diluted *Reference*-case and t is the residence time.

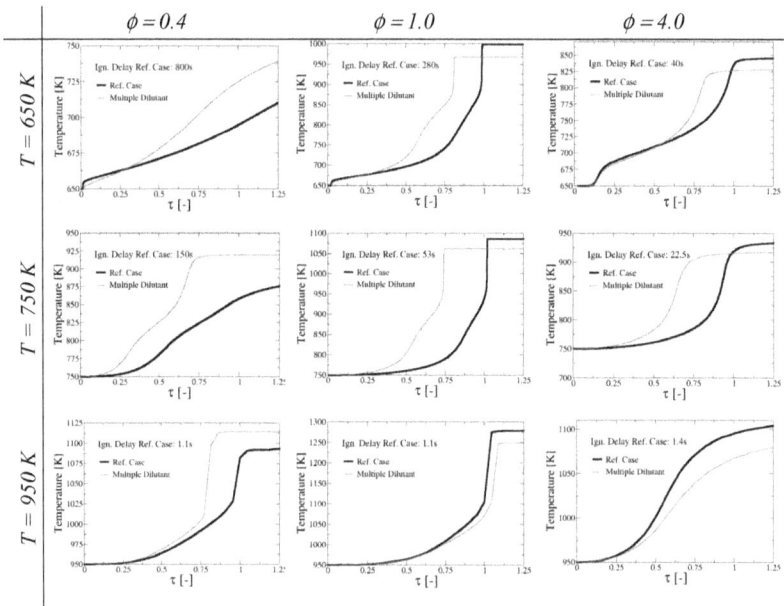

Figure 5-2   Temperature evolution during hydrocarbon oxidation diluted by different residual types at varying temperatures T and equivalence ratios ϕ (x_{res} = 0.97, P = 1 bar).

Compared to the *Reference*-case, the *Multiple-Dilutant*-case shows for all stoichiometric and rich conditions a decrease of the maximum temperatures obtained when reaction equilibrium is reached. According to Figure 2-4 in section 2.2, the specific heat capacities of CO_2 and H_2O are considerably larger than that of N_2 and consequently replacing N_2 by CO_2 or H_2O results in an increased global specific heat capacity. Thus, the maximum temperatures obtained for the *Multiple-Dilutant*-case are lower compared to those found for the *Reference*-case (disfavouring C_p-effect). However, this is not the case during lean combustion ($\phi = 0.4$), where the maximum temperatures obtained for the *Multiple-Dilutant*-case are greater than those found for pure N_2-dilution. The higher equilibrium temperatures under lean conditions result from the additional heat released due to the combustion of added CO. Under these conditions, CO reacts strongly via reactions (102) and (103) generating supplemental combustion heat.

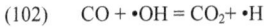

$$(102) \quad CO + \cdot OH = CO_2 + \cdot H$$

$$(103) \quad CO + \cdot OOH = CO_2 + \cdot OH$$

Figure 5-2 also reveals that, at temperatures below 950 K, the presence of CO, CO_2 and H_2O drastically shortens ignition delays compared to pure N_2-dilution, despite what might be expected from their increased heat capacities. This observation contradicts the general assumption that CO_2 and H_2O are chemically inert species and are supposed to act kinetically similarly to N_2. The promoting effect due to the presence of CO, CO_2 and H_2O is particularly important at intermediate initial temperatures (750 K).

Figure 5-3 compares the temperature evolutions during fuel oxidation at different dilution ratios x_{res} (0.86, 0.91 and 0.97 at $P = 1$ atm) and different initial pressures P (1, 10 and 20 bar at $x_{res} = 0.97$) for the equivalence ratios $\phi = 0.4$, 1.0 and 4.0. The tested initial temperature was 750 K.

The impact of the dilutant composition on mixture reactivity changes slightly with the total dilution ratio. For every dilution ratio, the mixture diluted by N_2, CO, CO_2 and H_2O (*Multiple-Dilutant*-case) is more reactive than the same mixture diluted by pure N_2 (*Reference*-case). Generally the mixture reactivity decreases with increasing dilution ratio and ignition delays are more or less doubled when the dilution x_{res} increases from 0.86 ($x_{EGR} = 0.52$) to 0.91 ($x_{EGR} = 0.6$). A further increase of the dilution ratio up to $x_{res} = 0.97$ ($x_{EGR} = 0.86$) leads to ignition delay times multiplied by 5 to 6 compared to a dilution of $x_{res} = 0.91$ ($x_{EGR} = 0.6$)

When pressure is changed under every dilution condition (*Reference*-case, *Multiple-Dilutant*-case) and equivalence ratio, the mixture reactivity increases with increasing pressure. At pressures close to the atmospheric pressure, HC oxidation is very sensitive to pressure changes. When the pressure rises from 1 bar to 10 bar ignition delay times are divided by around 10. However, a pressure rise from 10 to 20 bar results in quite lower kinetic acceleration and ignition delay times are divided by 3 under lean and rich conditions and 1.5 at stoichiometry. Changes in the dilutant composition reveal the strongest impact at atmospheric pressure. The impact of the dilutant composition is reduced with increasing pressure. However, at a pressure of 20 bar accelerated kinetics are observed as well for the *Multiple-Dilutant*-case.

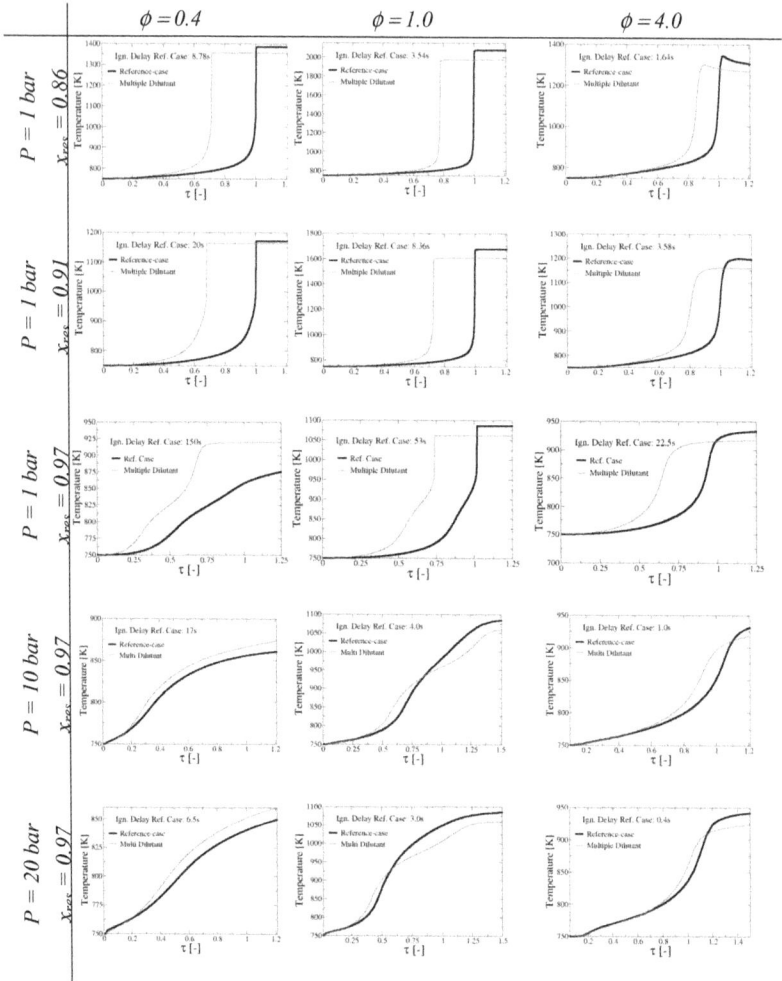

Figure 5-3   Temperature evolution during hydrocarbon oxidation diluted by different residual types at varying dilution ratios x_{res}, pressures P and equivalence ratios ϕ ($T_0 = 750$ K, P = 1 bar).

Single dilutant-Cases

In order to verify the thermal and kinetic impact of each isolated residual species, we replaced the N_2 of the *Reference*-case by 100 % of each investigated diluting compound (Table 18: *Single-Dilutant*-cases). Figure 5-4 displays the calculated temperature evolutions and compares them to the temperature evolution obtained for the *Reference*-case. Temperatures are plotted against the normalized time τ, which was introduced before.

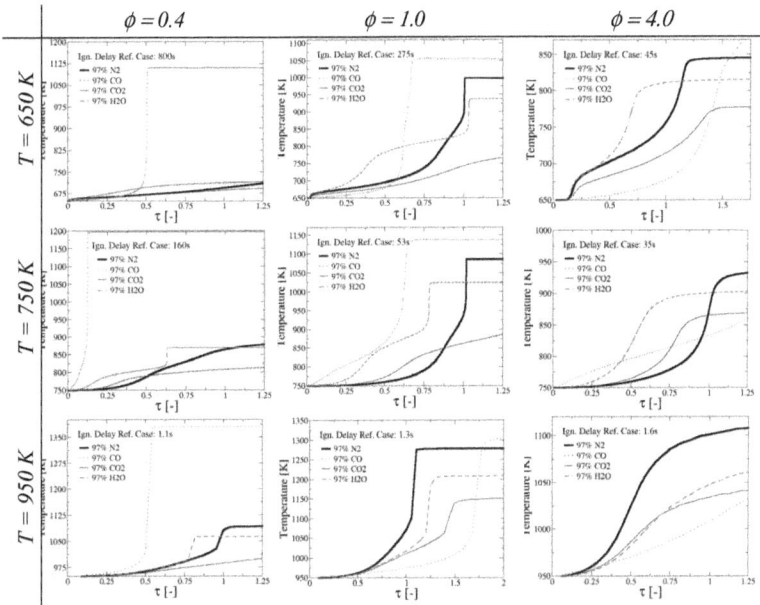

Figure 5-4   Temperature evolutions during hydrocarbon oxidation diluted by pure N_2, CO, CO_2 or H_2O at a dilution of 97 %, various equivalence ratios ϕ and at temperatures of 650, 750 and 950 K (x_{res} = 0.97, P = 1 bar).

Figure 5-4 shows strong variations in the adiabatic temperatures reached close to reaction equilibrium. Generally, the lowest maximum temperatures are obtained in the case of CO_2-dilution and the highest in case of CO. H_2O-dilution results in equilibrium temperatures ranging between those found for N_2 and CO_2. These observations correspond to the relative values of the specific heat capacities C_p shown in Figure 2-4 in section 2.2 for the considered species.

We mentioned already that the maximum temperatures reached for CO addition under lean conditions ($\phi < 0.4$) considerably overcome those obtained for pure N_2-dilution, due to the combustion heat generated from CO oxidation by excess oxygen. However, we also observe increased equilibrium temperatures at stoichiometric and rich conditions ($\phi = 1$, $\phi = 4$), resulting from N_2-substitution by CO. Such increase is due to the competition of fuel with the added CO for oxygen consumption. Direct CO oxidation generates more reaction heat than the competing incomplete fuel combustion which produces partially oxidized, unsaturated hydrocarbons. Thus, in the case of supplement CO, equilibrium temperatures are greater than those obtained for pure N_2-dilution. We also observed that at temperatures below 950 K, CO addition generally shortens main ignition delay times compared to N_2-dilution.

For the case of complete substitution of N_2 by CO_2 we observed at 750 K for all studied equivalence ratios a decrease of the residence time at which the mixture starts to react. However, except in rich conditions, the residence times for which equilibrium temperatures are reached are considerably greater compared to pure N_2-dilution. At 650 K and 950 K the complete substitution of N_2 by CO_2 generally results in increased ignition delays.

Similar to CO_2-dilution, at 750 K dilution by H_2O shortens the time when the mixture starts to react. In contrast to the case of CO_2, H_2O dilution results as well in reduced ignition delays and such behaviour is also observed at 650 K. This behaviour is noteworthy, as the C_p of H_2O is greater than that of N_2 and thus, an inhibiting impact on mixture reactivity might be expected. However, the increased reactivity by H_2O-dilution compared to the dilution by N_2 agrees with the observations made by Koroll and Mulpuru (1986), who found accelerated flame speeds when N_2 was replaced by H_2O.

The accelerated kinetics in the case of H_2O-dilution despite increased specific heat capacities suggests that the mixture reactivity is governed by a competition between 'disfavouring' thermal and 'favouring' kinetic effects. This competition is expected to depend strongly on the relative concentrations of different diluting compounds. At 950 K, the presence of CO, CO_2 and H_2O has an inhibiting kinetic effect in most cases.

In order to better understand the influence of the different diluting compounds, we tested their impact on the global reactivity in the concentration range in which they potentially may be observed during IC-engine postoxidation. The degree of dilution (97%) was kept constant and N_2 was gradually replaced by the different residual CO, CO_2 and H_2O (Table 18: Binary-Dilutant-cases) corresponding to the concentrations defined in the Multiple-Dilution-case. Figure 5-5 compares the obtained temperature evolutions for the following computed dilution cases:

		CO	CO_2,	H_2O	N_2
—	*Multiple-Dilutant*-case :	0.5 mol%	14 mol%	14 mol%	68,5 mol%
—	*Reference*-case :				97.0 mol%
---	*Binary-Dilutant*-case :	0.5 mol%			96.5 mol%
—	*Binary-Dilutant*-case :		14 mol%		83.0 mol%
–·	*Binary-Dilutant*-case :			14 mol%	83.0 mol%

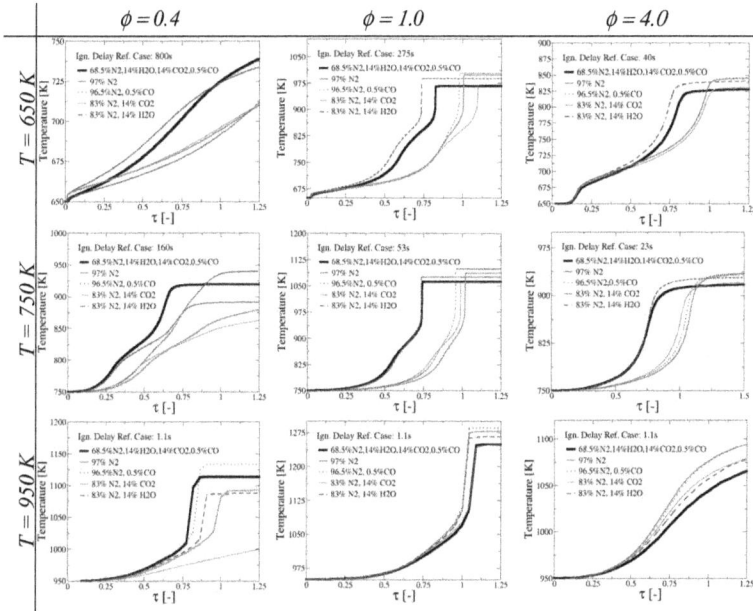

Figure 5-5   Temperature evolutions during hydrocarbon oxidation for different dilutant compositions at various equivalence ratios ϕ at temperatures of 650, 750 and 950 K and 97% dilution (x_{res} = 0.97, P = 1 bar).

At temperatures below 950 K, the *Multiple-Dilutant*-case is more reactive compared to the *Reference*-case, diluted totally by N_2. At 950 K, the *Reference*-case is the most reactive with exception of lean conditions (ϕ = 0.4). Like the complete replacement of N_2 (Figure 5-4), the partial N_2 substitution by CO, CO_2 or H_2O leads to different equilibrium temperatures, depending on the global reaction heat and specific heat capacities and due to the calorific power of CO, higher temperatures are reached when N_2 is partially replaced by CO.

At a temperature of 750 K, any partial N_2 substitution results in shorter ignition delays, even in the case of CO_2. In contrast to the complete N_2-replacement by CO_2 (Figure 5-4), its partial substitution by only 14% CO_2 results not only in reduced times at which the mixture starts to react, but as well in reduced ignition delays. However, the increased C_p-values of CO_2 compared to N_2 might suggest reduced temperature gradients during reaction, as it was observed in the case of 100% N_2-substitution by CO_2. We thus suppose a competition between '*disfavouring*' thermal and '*favouring*' kinetic influences of CO_2 on hydrocarbon oxidation. This competition seems to depend strongly on the CO_2-concentration-levels.

The strongest accelerating impact results from the partial N_2-substitution by H_2O. At temperatures below 950 K, the partial H_2O-dilution accelerates auto-ignition at almost any equivalence ratio. Thus, we conclude that the disfavouring C_p-effect of H_2O on fuel reactivity is suppressed by a favouring kinetic impact. The same should be true for the case of N_2-substitution by CO_2 at 750 K. However, in the case of H_2O, such favouring kinetic impact is much stronger than in the case of CO_2. At stoichiometric and rich conditions ($\phi = 1$, $\phi = 4.0$), the N_2/H_2O dilutant shows the same reactivity as the *Multiple-dilutant*-case.

5.1.2. Thermal versus kinetic impact: The cases of CO, CO_2 and H_2O

The purpose of this section is to analyse separately the thermal and kinetic impacts of the different diluting compounds. In order to quantify the differences between both effects, we have analysed separately the roles of the specific heat capacities and kinetic activities for each residual species. For that purpose, we have defined the following fictive species in the kinetic reaction mechanism which correspond to the investigated residuals N_2, CO, CO_2 and H_2O:

1. A fictive species corresponding to residual 'Y' is named 'AY'. It has the same thermal properties as 'Y', but collision efficiency factors of N_2. It is not involved in any reaction.

2. A fictive species corresponding to residual 'Y' is named 'BY'. It has the same thermal properties as 'Y' and also the same collision efficiency factors as 'Y'. It is however not involved in any reaction.

The concentrations of CO, CO_2 and H_2O defined in *Binary-Dilutant*-cases listed in Table 18 were replaced by their corresponding fictive species and simulations were performed. All calculations shown were run under stoichiometric conditions and at a constant temperature of 750 K. The concentrations of CO, CO_2 and H_2O defined in *Binary-Dilutant*-cases listed in Table 18 were replaced by its corresponding fictive species and simulations were performed. Figure 5-6 shows the calculated temperature evolutions plotted over the residence time t and compares them to those obtained for the *Reference*-case and the corresponding *Binary-Dilutant*-case (Table 18).

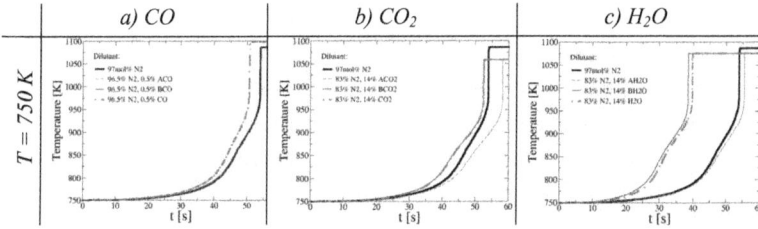

Figure 5-6 Temperature evolutions during stoichiometric hydrocarbon oxidation for the cases of partial-dilution by a) CO, b) CO₂ and c) H₂O for partial dilution by corresponding species 'Y', 'AY' and 'BY' compared to pure N₂-dilution. Thick black lines ▬ correspond to the Reference-case (pure N₂), thin dotted green lines --- to partial dilution by species 'AY', thin blue lines - to partial dilution by species 'BY' and thin red lines --- to partial dilution by species 'Y'.

Figure 5-6a shows the simulation results in the case of CO. The addition of the fictive species ACO and BCO leads to the same temperature evolutions as in the case of pure N_2-dilution. Thus, *the interactions of CO on hydrocarbon oxidation result exclusively from a direct participation of CO in chemical reactions.* Its thermal effect and its impact as collision partner are negligible.

Figure 5-6b presents the results for the corresponding cases of partial dilution by CO_2 and its corresponding fictive species (ACO_2 and BCO_2). Contrary to the addition of CO_2, the addition of fictive species ACO_2 considerably retards the occurrence of auto-ignition due to its large heat capacity. The ignition delay times for the case of fictive species BCO_2 is very close to that of CO_2, *showing that the major chemical influence of CO_2 results from its 3rd-body collision properties.* Its kinetic impact on the different kinetic reactions is therefore small at the investigated conditions.

Figure 5-6c displays temperature evolutions resulting from simulations for the partial dilution by H_2O performed with the corresponding fictive species (AH_2O and BH_2O). Very similar conclusions compared with the CO_2 case can be drawn with however a smaller thermal effect and a *larger influence as 3rd-body collision partner.* Due to its larger 3rd-body efficiency and its smaller disfavouring thermal impact H_2O, globally, has a much larger promoting effect than CO_2. Similar observations were made for different equivalence ratios.

5.1.3. Sensitivity Analysis

The investigation of thermal and kinetic impact of CO, CO_2 and H_2O on hydrocarbon reaction kinetics revealed that the influence of each compound on reactivity depends on its thermal properties (heat capacity), its kinetic characteristics (reactive compound, 3rd-body collision partner) and also on its concentration in the reaction. In order to quantify the impact of each residual species on hydrocarbon oxidation, we performed a sensitivity analysis under postoxidation conditions, at concentration levels at which the different diluting compounds may be found in IC-engine exhaust gases. That means we doubled separately the concentrations of each compound in the *Multiple-Dilutant*-case and compared the ignition delay times to those obtained for the *Multiple-Dilutant*-case. In that way, a sensitivity factor S_i was defined for each compound as follows:

$$[7] \quad S_i = \frac{t_{AI(Yi)} - t_{AI(2*Yi)}}{t_{AI(Yi)}}$$

where $t_{AI(2xYi)}$ is the ignition delay calculated for a doubled initial concentration of species y_i, and $t_{AI(Yi)}$ is the ignition delay obtained for the *Multiple-Dilutant*-case.

A positive sensitivity coefficient S_i indicates accelerated reaction kinetics as a result from doubling y_i's initial concentrations. Inhibited reaction kinetics are indicated by a negative sensitivity coefficient. Sensitivity analysis was performed for three equivalence ratios ($\phi = 0.4$, 1.0 and 4.0) at 750 K. In order to confirm our observations, we performed this analysis with two different detailed reaction mechanisms. We tested our reaction model described in chapter 4 and compared it to the detailed reaction mechanism proposed by Naik et al. (2005b). The later model was developed at the Lawrence Livermore National Laboratory (LLNL) and allows simulations of PRF/toluene oxidations. Figure 5-7 compares the sensitivity coefficients obtained from both mechanisms calculated for partial CO, CO_2 and H_2O-dilution at different equivalent ratios ϕ.

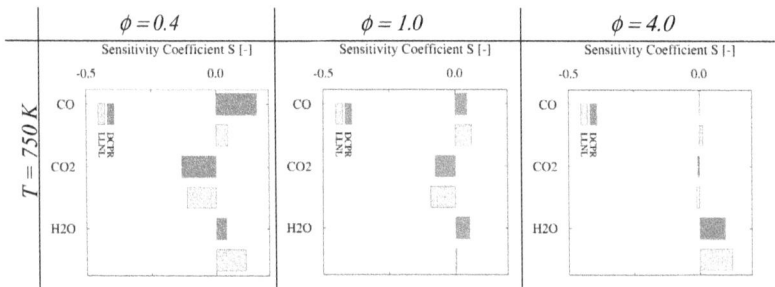

Figure 5-7   Sensitivity coefficients S obtained for CO, CO$_2$ and H$_2$O for varying thermodynamic conditions (T, ϕ) and different kinetic models: Green bars: Our modified reaction mechanism, blue bars Naik et al. (2005b).

Figure 5-7 shows that both mechanisms lead to similar sensitivity signs although their construction is different. Both mechanisms predict positive sensitivity coefficients in case of CO concentration variations under lean and stoichiometric conditions. This indicates accelerated kinetics due to increased CO concentrations. Under rich conditions ($\phi = 4$), the global reactivity is rather insensitive to CO addition. Both mechanisms confirm these observations.

Increasing CO_2 concentrations generally lead to inhibited reaction kinetics. This is in contrast to what was observed for partial substitution of N_2 by CO_2 for the same concentrations (14 mol%) and temperatures (Figure 5-5). Under similar conditions, the partial N_2-substitution by CO_2 results in accelerated kinetics (compare Figure 5-5). The accelerated kinetics resulting from partial N_2-substitution by CO_2 in a binary dilutant, but negative sensitivity (inhibition) when CO_2 is added in a realistic dilutant (dilution by N_2, CO, CO_2 and H_2O) might be explained by the competition between CO_2 and H_2O as 3rd-Body reaction collision partners. The collision efficiency of H_2O is greater than that of CO_2 and thus, in presence of H_2O, the accelerating impact of CO_2 as a 3rd-Body collision partner is lowered. The disfavouring thermal impact of CO_2 then becomes dominant and a net-inhibition of global reactivity resulting from increasing CO_2 concentrations is observed. Both tested mechanisms agree with that kinetic behaviour.

In the case of H_2O, accelerated reaction kinetics are observed for increasing H_2O concentrations at any stoichiometric condition. Our mechanism shows increasing sensitivity with increasing stoichiometry. This is also the case for the model of Naik et al. (2005b) except for stoichiometric conditions, where it is rather insensitive to an increase of H_2O concentrations. However, it should be outlined that our reaction mechanism as well that of Naik et al. (2005b) generally agree on the impact of the tested diluting species on hydrocarbon oxidation.

5.1.4. Flux Rate Analysis

Flux rate analysis was carried out in order to better understand the observed kinetic impact of CO_2 and H_2O as collision partners in 3rd-Body reactions. The performed rate analysis were based on our reaction model. Analysis was performed during stoichiometric fuel oxidation for pure N_2-dilution (*Reference-Case*), partial dilution by CO_2 and partial dilution by H_2O (corresponding *Binary-Dilution*-cases in Table 18). The reaction rates were analyzed at an initial temperature of 750 K. In order to compare the different cases to each other, we defined oxidation progress values for which the rate analysis was carried out. This means that flux rate analysis was performed for a reaction progress value, when a defined temperature of interest $T_{Flux\_Analysis}$ was reached. However, this results in different reaction times τ_i for each dilutant case i. Figure 5-8 shows the reaction times τ_i corresponding to the temperature of interest $T_{Flux\_Analysis}$. The choice of a temperature of interest $T_{Flux\_Analysis}$ allows the study of the different cases at comparable reaction progresses, despite variations in reaction time τ_i. Figure 5-9 shows the results of the performed reaction rate analysis.

$$\phi = 1.0$$

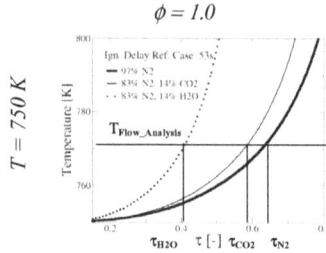

Figure 5-8 Temperature $T_{Flux\_Analysis}$ chosen for Rate Analysis and the corresponding reaction times τ_i.

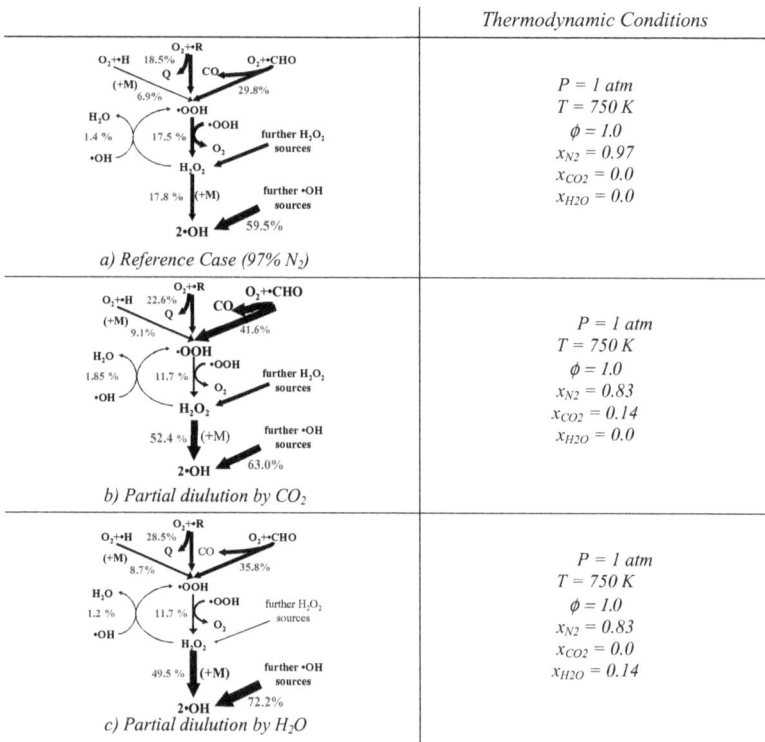

	Thermodynamic Conditions
a) Reference Case (97% N_2)	$P = 1\ atm$ $T = 750\ K$ $\phi = 1.0$ $x_{N2} = 0.97$ $x_{CO2} = 0.0$ $x_{H2O} = 0.0$
b) Partial diulution by CO_2	$P = 1\ atm$ $T = 750\ K$ $\phi = 1.0$ $x_{N2} = 0.83$ $x_{CO2} = 0.14$ $x_{H2O} = 0.0$
c) Partial diulution by H_2O	$P = 1\ atm$ $T = 750\ K$ $\phi = 1.0$ $x_{N2} = 0.83$ $x_{CO2} = 0.0$ $x_{H2O} = 0.14$

Figure 5-9 Fluxes of •OOH and H_2O for : a) the Reference-case, and the Binary-Dilution cases of b) CO and c) CO_2.

The fluxes indicated in Figure 5-9 are normalized by the consumption of oxygen calculated for the *Reference*-case at the corresponding temperature $T_{Flux\_Analysis}$. A strong increase of H_2O_2 reaction flow is observed when N_2 is partially substituted by CO_2 and H_2O.

The investigated temperature of 750 K is in the regime where cool-flames are observed and when the combustion intermediate H_2O_2 and hydroperoxy radicals •OOH govern hydrocarbon oxidation kinetics (Dagaut et al. 1995, Pilling et al. 1997, Westbrook 2000). In these conditions •OOH radicals are primary produced by the oxidation of •CHO to CO by reaction (104):

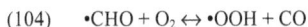

$$(104) \quad •CHO + O_2 \leftrightarrow •OOH + CO$$

and by the 3rd-Body reaction of O_2 and •H radicals to give •OOH (reaction (105)).

$$(105) \quad O_2 + •H (+M) \leftrightarrow •OOH \ (+M)$$

Hydroperoxy radicals •OOH are the major source of H_2O_2 which is generally produced from the recombination of •OOH radicals by reaction (106):

$$(106) \quad •OOH + •OOH \leftrightarrow H_2O_2 + O_2$$

During cool flame oxidation, H_2O_2 is mainly consumed by its dissociation via the 3rd-Body reaction (107):

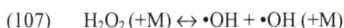

$$(107) \quad H_2O_2 (+M) \leftrightarrow •OH + •OH (+M)$$

which competes with the kinetically inhibitive •OH consuming reaction (108):

$$(108) \quad H_2O_2 + •OH \leftrightarrow •OOH + H_2O$$

Reaction (107) has an accelerating impact on hydrocarbon oxidation, while reaction (108) inhibits global reactivity. During cool flame oxidation, the consumption of H_2O_2 by reactions (107) and (108) is smaller than its production from •OOH. H_2O_2 is accumulated and its rate of consumption governs the global reaction kinetics.

At 750 K the rate of H_2O_2-dissociation is almost tripled, when N_2 is partially substituted by CO_2 or H_2O. The observed key role of H_2O_2-dissociation on global reactivity agrees with the conclusions drawn by Westbrook, (2000) who analyzed chemical kinetic factors of hydrocarbon oxidation in a variety of ignition problems. Among others, he studied the case of HCCI-engine control and identified the decomposition of H_2O_2 as major chain branching step during hydrocarbon oxidation at intermediate temperatures. Thus, he concluded that HCCI-ignition is controlled by H_2O_2-decomposition.

5.1.5. Kinetics of H_2O_2 decomposition

The performed flux rate analysis revealed the importance of H_2O_2 on chemical kinetics at low and medium temperature (600 K–800 K). It was shown that H_2O_2-dissociation (reaction (107)) governs global reactivity. Both tested mechanisms showed qualitatively comparable impacts of CO, CO_2 and H_2O on hydrocarbon oxidation. However, we observed differences in their sensitivities on CO, CO_2 and H_2O addition. We therefore compared the kinetics of H_2O_2-dissociation defined in both mechanisms.

According to Hughes et al. (2001), in the case of H_2O_2-dissociation, the greatest collision efficiency is attributed to H_2O (6.5) followed by CO_2 (1.5), while that of N_2 (0.4) is small in comparison. These values have been used in the mechanism of Anderlohr et al. (2009b). This results in remarkable changes in the reaction rate of H_2O_2-dissociation when the mixture is diluted by CO_2 and H_2O (*Multiple Dilutant*-case) in comparison to pure N_2 dilution (*Reference*-case). Figure 5-10 shows the reaction rate of H_2O_2-dissociation as a function of temperature. Reaction rates were computed for the reaction mechanisms proposed by Anderlohr et al. (2009b) and Naik et al. (2005) and compared for the dilution conditions defined for the *Multiple Dilutant*-case and for the *Reference*-case conditions.

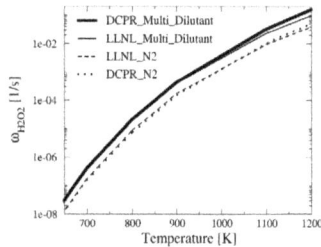

Figure 5-10 Reaction rate of H_2O_2-dissociation as a function of temperature computed under Multiple Dilutant-case and Reference-case conditions for the developed reaction mechanism and that proposed by and Naik et al. (2005).

Our reaction mechanism as well as that of Naik et al. (2005) show a significant change of the H_2O_2-dissociation reaction rate when CO_2 and H_2O are present. Both mechanisms show an increase of the H_2O_2-dissociation reaction rate under the dilution conditions of the *Multiple Dilutant*-case. Table 19 shows the rate constants and collision efficiencies for H_2O_2-dissociation defined in both mechanisms.

Table 19   *Kinetics of H_2O_2-dissociation defined in the mechanism of Anderlohr et al. (2009b) and Naik et al. (2005b).*

Reactions		A	n	E_a	References	No
•OH+•OH+M→ H_2O_2+M	*(high pressure)*	7.23×10^{13}	-0.37	0.0	*Present work*	(107) sens 1
	(low pressure)	5.53×10^{19}	-0.76	0.0		
H_2O_2+M→ •OH+•OH+M	*(high pressure)*	3.0×10^{14}	0.0	48500	*Present work*	(107) sens (-1)
	(low pressure)	3.0×10^{17}	0.0	45500		
	Fall off parameter :	Troe /0.5/1/1x10^8				
	Collision efficiencies	N_2: 0.4, O_2: 0.4, H_2O: 6.5, CO: 0.75, CO_2: 1.5				
•OH+•OH+M ↔ H_2O_2+M	*(high pressure)*	1.236×10^{14}	-0.37	0.0	*Naik et al. (2005b)*	(107)
	(low pressure)	3.041×10^{30}	-4.63	2049		
	Fall off parameter :	Troe /0.47/100/2000/1x10^{15}				
	Collision efficiencies	N_2: 1.0, O_2: 1.0, H_2O: 12, CO: 1.9, CO_2: 3.8				

The rate constants are given ($k=A\ T^n\ exp(-E_a\ /RT)$) in cc, mol, s, cal units.

The reaction rate of H_2O_2 dissociation to •OH radicals (reaction (107)) is strongly dependent on the characteristics of its collision partners (+M). The collision efficiency of CO_2 and H_2O differ strongly from that of N_2 (Brabbs et al.1987, Baulch et al. 1992, 1994, 2005). In the case of H_2O_2-dissociation the greatest collision efficiency is attributed to H_2O followed by CO_2, while that of N_2 is small in comparison (Baulch et al. 1992, 1994, 2005). Thus, the observation of accelerated •OH production resulting from N_2-substitution by CO_2 and H_2O at 750 K is reasonable. Our reaction model defines H_2O_2-dissociation by two reactions, a forward and a backward reaction. Naik et al. (2005b) consider one single equilibrium reaction. Comparing collision efficiencies we observe strong differences in the efficiency of the defined species. However, both mechanisms consider similar ratios of H_2O, CO_2 and CO efficiencies compared to the N_2 and O_2 efficiencies.

In order to test the impact of the different H_2O_2 kinetics we defined in our reaction model the H_2O_2-dissociation identically to those of the mechanism of Naik et al. (2005b). With the obtained modified mechanism we performed simulations for the partial N_2-substitution by CO, CO_2 and H_2O at stoichiometric conditions and at a temperature of 750 K. Figure 5-11 shows the temperature evolutions calculated with the modified mechanism and compares them to those calculated with our original mechanism.

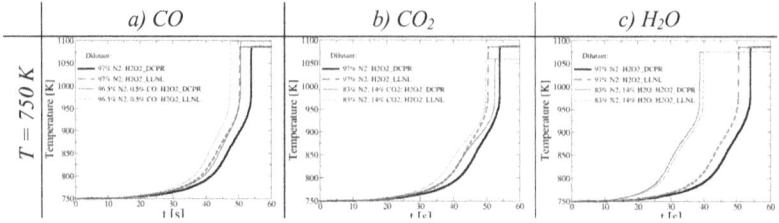

Figure 5-11 Temperature evolutions during stoichiometric hydrocarbon oxidation for the Reference-case and the cases of partial-dilution by a) CO, b) CO₂ and c) H₂O computed with the reaction kinetics of H₂O₂ dissociation applied by Naik et al. 2005 (LLNL) and in the presented reaction mechanism (DCPR).

The simulations shown in Figure 5-11 reveal that in any case of partial N_2-substitution the kinetics of H_2O_2-dissociation defined in the mechanism of Naik et al. (2005b) accelerate considerably the global reactivity. However, the qualitative temperature evolution profiles remain the same.

Our investigation shows the strong sensitivity of low temperature hydrocarbon oxidation on H_2O_2-dissociation. H_2O_2-dissociation is strongly governed by the efficiencies of its 3rd-body collision partners. However, estimations of collision efficiencies differ strongly in various detailed reaction mechanisms and thus, improved estimations of 3rd-body collision efficiencies are necessary.

5.1.6. Concluding remark on the impact of residual gases on HC oxidation

The thermal and kinetic impact of the residuals CO, CO_2 and H_2O on hydrocarbon oxidation were investigated numerically. The case of dilution by N_2 was tested against a dilutant composed of CO, CO_2 and H_2O in proportions found at IC-engine postoxidation conditions. The influence of CO on hydrocarbon oxidation is restricted to its direct participation on oxidizing reactions and its thermal impact is negligible compared to N_2. CO_2 and H_2O have a negative thermal impact compared to N_2 resulting from their increased heat capacity. Kinetically, they interact with hydrocarbon oxidation through H_2O_2-dissociation as collision partners. Generally, it was shown that the composition of a dilutant strongly impacts hydrocarbon oxidation and that the presence of CO_2 and mainly H_2O may lead to an acceleration of hydrocarbon oxidation, in spite of their unfavourable thermal properties compared to N_2.

Based on the developed reaction mechanism an analysis was performed for various dilutant proportions, equivalence ratios and temperatures. The impact of each residual was tested individually, as well as in combination with the others. Strongly accelerated kinetics were observed when pure N_2-dilution was replaced by a [N_2, CO, CO_2 and H_2O] dilutant.

Table 20 indicates typical proportions of the investigated species in engine exhaust gases and their general impact on hydrocarbon oxidation observed under postoxidation conditions ($T < 900$ K, high degree of dilution).

Table 20   General impact of residual species on hydrocarbon oxidation kinetics at temperatures below 900 K.

Species	Proportion. in the dilutant [mol/mol]	General impact on reaction kinetics
N_2	0.685	neutral (Rreference)
CO	0.005	accelerating
CO_2	0.14	quasi neutral
H_2O	0.14	accelerating

Sjöberg et al. (2007) experimentally investigated the impact of EGR diluting compounds on HCCI-engine control. They tested different types of EGR and distinguished the impact of the individual EGR constituents N_2, CO_2 and H_2O and found an accelerating impact on ignition delays resulting from the presence of H_2O. The simulation results presented in this chapter agree well with the observations made by Sjöberg et al. (2007).

It was seen that a binary H_2O/N_2-dilution kinetically behaves very similarly to the corresponding N_2, CO, CO_2 and H_2O dilution (same H_2O-concentrations). By defining fictive species in the reaction mechanism it was possible to analyse separately thermal and kinetic effects. It was shown that an accelerating impact results from the addition of CO due to its calorific power and oxidation potential. In case of CO_2 and H_2O, a negative thermal C_p-effect competing with an accelerating kinetic impact due to 3rd-Body reactions was observed.

Flux rate analysis revealed the importance of H_2O_2-dissociation on global reactivity at low and intermediate temperatures. It was shown that H_2O_2-dissociation is strongly governed by the efficiencies of its collision partners. Significant uncertainties exist on collision efficiencies of various potential collision partners. However this information is crucial for the comprehension and modelling of low temperature oxidations. Thus, there is a strong need of further experimental validation of 3rd-body reactions and the efficiencies of potential collision partners.

5.2   Study on the formation of C_1-C_2 species

This section presents a numerical study of the impact of various additives on the oxidation of a typical automotive surrogate fuel blend, i.e. n-heptane and toluene based on experiments that were performed by Piperel (Anderlohr et al. 2009b) at the ICARE laboratory in Orléans. It examines the impact of engine re-cycled exhaust gas compounds on the control of an HCCI-engine. Series of experiments were performed in a highly diluted jet-stirred reactor at pressures of 1 and 10 atm. The chosen thermo-chemical conditions were close to those characteristic of the pre-ignition period in an HCCI-engine. The influence of various additives, namely nitric oxide (NO), ethylene and methanol, on the oxidation of a n-heptane/toluene blend was studied over a wide range of temperatures (550-1100 K), including the zone of the Negative Temperature Coefficient (NTC).

A new detailed chemical kinetic reaction mechanisms based on that described in chapter 4, is proposed for the oxidation of the surrogate fuel. It includes reactions of NO, methanol and ethylene and is used to explain the obtained experimental data. The mechanism is further used to theoretically study the impact of other EGR compounds on the hydrocarbon oxidation, namely ethane, formaldehyde and carbon monoxide.

5.2.1. Optimization of reactions producing C_1-C_2 species

The standard EXGAS-mechanisms tend to overpredict ethylene concentrations at low temperatures. Since the modelling of ethylene concentrations is of major importance for this study, the ethylene producing reactions, defined in the secondary mechanism, were improved. The applied modifications concern the decompositions of hydroperoxides and cyclic ethers, as well as the H-abstractions by •O, •OH, •OOH and methyl radicals from olefins. This promotes the formation of larger alkyl or alkenyl radicals, instead of producing small C_0-C_2 radicals in one-step reactions.

We thus revised the secondary mechanism of the EXGAS standard version and redefined the decompositions of hydroperoxides (ROOH), ketohydroperoxides (QOOOH), cyclic ethers and cyclic-ether-keto –hydroperoxides. We redefined as well reactions of olefins and ketones. Our modelling strategy was not to decompose such species directly to C_1-C_2 species, but to decompose them when possible to larger species (C_n, n > 2) which are re-injected into the primary mechanism. For better reproducing methanol profiles we added also the disproportionation between methylperoxy and other alkylperoxy radicals, leading to the formation of oxygen, aldehyde and methanol molecules, as proposed by

Lightfoot et al. (1992). The corresponding reactions were added to the primary mechanism and allowed to explain the observed formation of methanol around 650 K.

All reaction types which were added or modified in the EXGAS-Standard version for this study are listed in Table 21. Exemplary for all alkanes and isomers for each reaction type, the reactions of one representative C_7 isomer is shown. For a better understanding the modified reactions are compared to the reactions originally written in the EXGAS-standard version. Reactions which were added to the EXGAS-standard version are marked in the table as '*added*', reactions which were modified as '*modified*' and the reactions written as in the EXGAS-standard version are referenced as '*original*'. The reactions originally defined in the secondary mechanism in the EXGAS-standard version consider the decomposition of large oxygenated molecules (C_n, n > 2) to small molecules of the C_0-C_2 reaction base.

One major product of such decompositions is ethylene (see Table 21: reactions referenced '*original*'). Those reactions defined in the secondary mechanism result in a strong production of ethylene at low to intermediate temperatures (600 K – 850 K) when alkanes are mainly decomposed via the addition of oxygen and the formation of hydroperoxy radicals. The revised decompositions defined in the secondary mechanism (see Table 21: reactions referenced '*modified*') decompose the oxygenated products issued from the primary mechanism, if possible, to molecules and radicals, which are part of the primary mechanism (re-injection of radicals of the secondary mechanism into the primary mechanism). This allows an iterative decomposition of large molecules into small species which are part of the C_0-C_2 reaction base and products of the secondary mechanism are not stringently injected directly to the C_0-C_2 reaction base. These improvements agree with those proposed by Biet et al. (2008).

We revised as well the production of formaldehyde defined in the secondary mechanism and found that the main sources of formaldehyde are the decomposition of cyclo-ether ketohydroperoxydes and the reaction of olefins with •OH radicals (modified reactions 115 and 116). The mechanism which was improved in the described way will be further referenced as modified EXGAS-version. Figure 5-12 compares the measured concentration-profiles of n-heptane, ethylene and methanol with the profiles calculated using n-heptane models generated by the standard EXGAS-version and by the modified EXGAS-version. .

The changes made in the secondary mechanism for the oxidation of n-heptane led to a significant improvement in the reproduction of species concentration-profiles. Despite an overestimation of n-heptane consumption below 800 K (Figure 5-12a), the reduction of fuel conversion with increasing temperature in the NTC-region is better predicted by the modified EXGAS-version. In addition, the calculated ethylene concentration-profile (Figure 5-12b) is closer to the experimental observations and the methanol concentration peak around 650 K is more accurately reproduced.

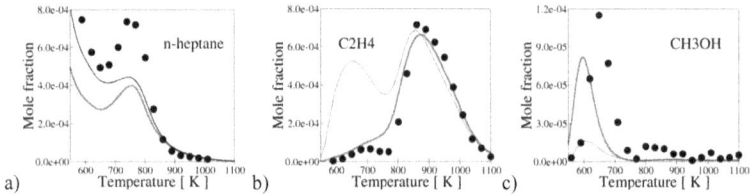

Figure 5-12 Experimental (symbols) and simulated (lines) mole fractions of (a) n-heptane, (b) ethylene and (c) methanol without additive. Simulations were performed with the modified EXGAS-version (thick red full lines —) and with the standard EXGAS-version (thin black full lines -) mechanisms.

The following section first discusses the comparison between experimental and simulated results in the presence of NO, ethylene and methanol (Figure 5-13) and analyzes the reactions leading to the observed effects. The model is then further used for predicting the impact of the addition of formaldehyde, ethane and CO on the fuel oxidation (Figure 5-14).

5.2.2. Model validation

Figure 5-13 presents the measured concentrations of different species as functions of temperature at an equivalence-ratio of 0.75 for pure fuel, a fuel/NO mixture and a mixture of fuel/ethylene/methanol. The initial concentrations of n-heptane and toluene were held constant at 900 and 100 ppm respectively, and the influence of the additives methanol (130 ppm), ethylene (200 ppm) and NO (100 ppm) was tested separately and together. Concerning the addition of NO, results are shown for mixtures of n-heptane/toluene with the addition of NO only.

We show here the results with addition of NO and with addition of both methanol and ethylene, as close results were obtained for these compounds added separately and interactions between methanol and ethylene additives were not observed.

Table 21   Modified reactions in the secondary mechanism compared to the EXGAS standard version.

Reactions	A	n	E_a	Reference	No
Formation of Methanol					
ROO• + ROO• → ROH +R'CHO + O_2	1.25×10^{10}	0	-8000	Estimated[a] *(added)*	109)
Hydroperoxide decomposition					
•OH + CH_3CHO +•C_5H_{11}	1.5×10^{16}	0	42000	KINGAS *(modified)*	110)
•OH+HCHO+2*C_2H_4+•C_2H_5	1.5×10^{16}	0	42000	KINGAS *(original)*	(112)
Ketohydroperoxide decomposition					
•OH+CH_3CHO+CO+•C_4H_9	1.5×10^{16}	0	42000	KINGAS *(modified)*	111)
•OH+HCHO+CO+2*C_2H_4+•CH_3	1.5×10^{16}	0	42000	KINGAS *(original)*	(113)
O-ring decomposition					
+ •OH ⟶ H_2O+CH_2CO+ C_2H_4+•C_3H_7	7.8×10^{6}	2	5000	KINGAS *(modified)*[b]	112)
+ •OH ⟶ H_2O+CH_2CO+ 2*C_2H_4+•CH_3	7.8×10^{6}	2	5000	KINGAS *(original)*[b]	(114)
Decomposition of cyclo-ether ketohydroperoxydes					
•OH+•C_2H_3+HCHO+ CH_3CHO+CH_2CO	1.5×10^{16}	0	42000	KINGAS *(modified)*	113)
•OH+CO_2+CH_3CHO+2*C_2H_4	1.5×10^{16}	0	42000	KINGAS *(original)*	(115)
Reaction of OH with olefines					
+ •OH ⟶ HCHO+C_2H_4+•C_4H_9	1.4×10^{12}	0	-900	KINGAS *(modified)*	114)
+ •OH ⟶ H_2O+C_2H_4+•C_5H_{11}	3.1×10^{6}	2	-300		115)
+ •OH ⟶ HCHO+•C_2H_5+2*C_2H_4	1.4×10^{12}	0	-900	KINGAS *(original)*	(116)
+ •OH ⟶ H_2O+C_4H_6+C_2H_4+•CH_3	3.1×10^{6}	2	-300		(117)
Ketones reactions					
CO +•OH ⟶ H_2O+ CH_2CO+ •C_5H_{11}	7.8×10^{6}	2	5000	KINGAS *(modified)*	116)
CO +•OH ⟶ H_2O+ CH_2CO+ 2*C_2H_4+•CH_3	7.8×10^{6}	2	5000	KINGAS *(original)*[b]	(118)

The rate constants are given ($k=A\ T^n \exp(-E_a/RT)$) in cc, mol, s, cal units.

[a]: The rate constant was equal to 10 times the value proposed by Lightfoot et al. (1992) for the reaction of CH_3OO• radicals.
[b]: The reaction rates of corresponding species with a different number of carbon atoms reacting with •OH, •H, •OOH, •CH_3, CH_3OO• and •C_2H_5 radicals were defined analogous. Reaction rates vary in function of the radical reacting with the concerned species.

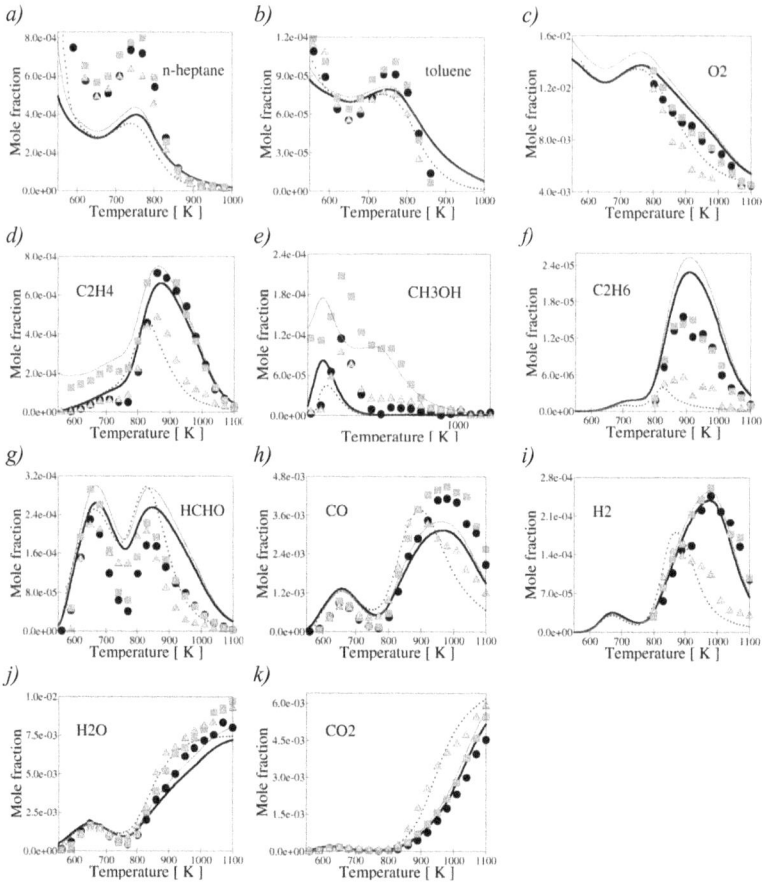

Figure 5-13 Experimental (symbols) and simulated (lines) mole fractions of (a) n-heptane, (b) toluene, (c) oxygen and (d-k) the main combustion intermediates with no additive (black circles • and thick full lines ▬), with addition of ethylene and methanol (green squares and thin lines) and with addition of NO (blue shaded triangles ⊿ and dotted lines ---).

Figure 5-13a (n-heptane) and Figure 5-13b (toluene) show that the addition of NO inhibits the oxidation of the fuel at low temperatures (<700 K), accelerates it at high temperatures (> 850 K) and reduces the intensity of the NTC. The accelerating effect above 800 K is also observed on the oxygen-concentration-profile (Figure 5-13c), showing that the addition of NO induces a strong depletion above 900 K. This result agrees with the observations of Dubreuil et al. (2007) and Moréac et al. (2006) in their studies on the complex impact of NO on HC oxidation. At temperatures below 850K, the addition of ethylene and methanol leads to a slightly retarded oxidation of the fuel, whereas at higher temperatures, this addition has no impact on the fuel oxidation. The modified EXGAS-version quantitatively reproduces well these observations. However, the consumption of n-heptane is overestimated (Figure 5-13a).

For temperatures below 850 K, the effect of NO on the ethylene concentration-profile (Figure 5-13d) is negligible. Above 850 K, the formation of ethylene is strongly reduced when NO is added and the ethylene peak-concentration, observed at 850 K, decreases strongly. Below 850 K, when ethylene is added, its measured concentrations are increased by the magnitude of the initially added amounts. At higher temperatures, ethylene-addition does not affect its measured concentration-profiles, which become quasi independent of the initially added ethylene concentration.

Figure 5-13e shows that generally, at around 650 K, a production peak of methanol from the fuel oxidation is observed. This production is reduced when NO is added. For temperatures above 700 K and in the case of methanol-addition, the methanol concentration-profile also includes a NTC-zone which is superposed by the mentioned methanol production peak. One observes that the ethane concentration (Figure 5-13f) is not greatly affected by the addition of ethylene and methanol, but is strongly reduced when NO is added. The presence of NO leads to an increase of the formaldehyde concentration in the NTC-region (700-800 K) and to a decrease at higher temperatures. This result is consistent with the observations made by Dagaut et al. (1995). Our model predictions agree very well with these observations and the modified EXGAS-version reproduces correctly concentrations of ethylene in presence of NO over the complete temperature range (Figure 5-13d), the production peak of methanol around 650 K (Figure 5-13e), as well as the reduced production of ethane in presence of NO (Figure 5-13f).

At temperatures below 850 K the addition of methanol and ethylene increases the measured formaldehyde concentration (Figure 5-13g), but at higher temperatures, the formaldehyde concentration-profile is unchanged by their addition. Ethylene and methanol-addition has little effect on the formation of CO (Figure 5-13h), H_2 (Figure 5-13i), H_2O (Figure 5-13j) and CO_2 (Figure 5-13k). At temperatures above 800 K, these additives slightly increase the concentrations of these oxidation products. In contrast, the addition of NO increases the CO and hydrogen production between 800 K and 900 K, and considerably decreases it at higher temperatures. Consequently, the formation of H_2O and CO_2 is also increased by the addition of NO at such temperatures. The predictions of the modified EXGAS-version agree very well with these observations.

5.2.3. Kinetic study of the impact of the addition of NO, C_2H_4 and CH_3OH

The following sections present the comparison of experimental data to model predictions.

5.2.3.1. Addition of NO

Figure 5-13a-c show that the complex impact of NO on the consumption of reactants is accurately predicted throughout the whole considered temperature range. The slight retarding effect of NO which is observed below 700 K is mainly caused by reaction type (33) of alkylperoxy radicals (ROO•) with NO to give alkoxy radicals:

$$(33) \qquad ROO•+NO \leftrightarrow RO•+NO_2$$

These reactions compete with the isomerization of ROO• radicals followed by a second addition of oxygen yielding branching agents (hydroperoxides). The NO promoting effect observed above 700 K is due to the reaction (31) between NO and •OOH radicals, transforming these rather unreactive radicals into •OH radicals. The following catalytic cycle can be written, with reaction (31) consuming NO and reactions (32), (53) and (64) regenerating it:

$$(31) \qquad NO+•OOH \leftrightarrow NO_2+•OH$$
$$(32) \qquad NO_2+•H \leftrightarrow NO+•OH$$
$$(53) \qquad NO_2+•CH_3 \leftrightarrow NO+CH_3O•$$
$$(64) \qquad NO_2+•C_2H_5 \leftrightarrow NO+C_2H_5O•$$

The effect of this cycle is a noticeable reduction of the concentration of •OOH radicals and an increase of •OH radicals. Figure 5-13d-k show that the effects of NO addition on the formation of the main products are correctly reproduced. In general, these effects are:

- An important reduction in the formation of ethylene above 800 K (Figure 5-13d) which is caused by reaction (64) becoming competitive with reaction (117), which is the main ethylene formation channel:

$$(117) \qquad •C_2H_5+O_2 \leftrightarrow •OOH+C_2H_4$$

- A decrease in the formation of methanol around 650 K (Figure 5-13e) due to the NO consumption of methylperoxy radicals by reaction (118), which are the main source of methanol via reaction (119):

$$(118) \qquad CH_3OO•+NO \leftrightarrow CH_3O•+NO_2$$
$$(119) \qquad ROO•+CH_3OO• \leftrightarrow RCHO+CH_3OH+O_2$$

- A lower production of ethane (Figure 5-13f) which is caused by reaction (53) consuming •CH₃ radicals and thus reducing the formation of ethane, obtained via reaction (120):

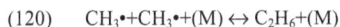

$$(120) \quad CH_3\bullet + CH_3\bullet + (M) \leftrightarrow C_2H_6 + (M)$$

- A complex impact on the formation of formaldehyde above 700 K (Figure 5-13g). As reactions (53) and (64) favour the formation of formaldehyde by an enhanced production of $CH_3O\bullet$ and $C_2H_5O\bullet$ radicals, the production of aldehydes increases due to the rapid decomposition of these radicals. This effect, enhanced by the increased global reactivity, explains the higher concentration of formaldehyde between 750 and 850 K resulting from the addition of NO. At higher temperatures, the increased formation of •OH radicals promotes the consumption of formaldehyde and explains its lower concentration in presence of NO.

- Effects of NO on the concentrations of CO (Figure 5-13h), H_2 (Figure 5-13i), H_2O (Figure 5-13j) and CO_2 (Figure 5-13k). The enhanced global reactivity induced by NO at temperatures above 650 K results in an increased formation of CO and H_2. However, above 900 K, the production of •OH radicals by reactions (31) and (32) becomes important and induces a larger consumption of CO thus producing a larger concentration of CO_2. As •OH radicals are consumed by formaldehyde producing •CHO radicals and H_2O, the production of water also increases. As the main source of H_2 is the H-abstractions by H-atoms from aldehydes, an increase of •OH production results in a H_2 concentration drop off due to the competition between reactions (121) and (122):

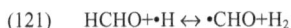

$$(121) \quad HCHO + \bullet H \leftrightarrow \bullet CHO + H_2$$

$$(122) \quad HCHO + \bullet OH \leftrightarrow \bullet CHO + H_2O$$

The favoured formation of CO_2 and water comes along with the increased consumption of oxygen above 900 K when NO is added.

5.2.3.2. Addition of ethylene and methanol

Figure 5-13a and b show that the addition of ethylene and methanol results in a reduced global reactivity below 800 K. This is correctly reproduced by the model. Ethylene and methanol are stable compounds which are mainly consumed by the attack of •H and •OH radicals resulting in less reactive radicals and therefore representing a non negligible sink for reactive radicals. This competition in the consumption of •H and •OH radicals between the additives (ethylene, methanol) and the initial fuel reduces the oxidation-rate of n-heptane and toluene. At higher temperatures, this competition becomes negligible compared to the rate of production of •H and •OH radicals. Therefore, the impact of these additives on the fuel oxidation becomes much smaller. For temperatures up to 1000 K the production of ethylene, an important intermediate product of hydrocarbons oxidation, exceeds its consumption (Figure 5-13d). Only at higher temperatures the ethylene consumption is so strong that its concentration becomes quasi-independent of the initially added amount of ethylene.

At temperatures around 600 K, the methanol concentration produced by reactions exceeds the amount of introduced methanol (Figure 5-13e). The observed concentration peak corresponds to the amount of methanol which is produced via reaction (119). At higher temperatures (> 700 K), the consumption of methanol exceeds its production and thus its concentration-profile corresponds to that observed for the oxidation of n-heptane and toluene which is characterized by a NTC-zone between 700 and 800 K. The presence of ethylene and methanol enhances the formation of formaldehyde through the depletion of •C$_2$H$_3$, •CH$_2$OH and •CH$_3$O radicals.

5.2.3.3. Study of the effect of the addition of C$_2$H$_6$, HCHO and CO

The good agreement between experimental and modelling results justifies using the model for predicting the impact of various organic additives on the fuel oxidation. Ethane, HCHO and CO are known to be present in EGR and may influence the HCCI-engine control. The model was then used to study the chemical impact of these species on the fuel oxidation.

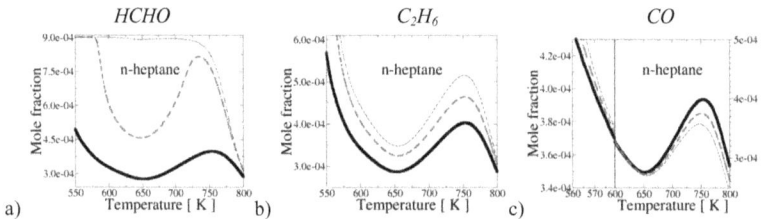

Figure 5-14 Simulated mole fractions of n-heptane with (a) formaldehyde-additive, (b) ethane-additive and c) CO-additive. Additive mole fractions are 0ppm (thick line —), 800ppm (green dotted line ---) and 1600ppm (blue dotted line ---).

Figure 5-14a compares the impact of various amounts of formaldehyde on the oxidation of n-heptane. The simulations show that the addition of formaldehyde has a significant retarding effect on the n-heptane oxidation at low temperatures (< 800 K) but that it also increases the NTC-amplitude. At high concentrations of added formaldehyde (1600 ppm), the resulting retarding effect becomes so strong that up to 750 K no reactivity of the mixture is predicted. Formaldehyde is consumed by the attack of •OH radicals via reaction (123), while the •CHO radicals react by reaction (124):

---

(123) $HCHO+\bullet OH \leftrightarrow \bullet CHO+H_2O$

(124) $\bullet CHO+O_2 \leftrightarrow CO+\bullet OOH$

The sum of reactions (123) and (124) results in a net-transformation of the reactive $\bullet OH$ into the less reactive $\bullet OOH$ radicals, thus explaining the increased NTC-amplitude and the lack of reactivity at temperatures below 800 K for high concentrations of added formaldehyde.

Figure 5-14b shows the impact of various amounts of ethane on the oxidation of n-heptane. The addition of ethane yields an inhibition at low temperatures, but also leads to a slight increase of the NTC-amplitude, which is less apparent than when formaldehyde is added. This amplification caused by ethane can be explained by the formation of ethyl radicals yielding an increased production of $\bullet OOH$ radicals via reaction (117).

Figure 5-14c compares the impact of various amounts of CO on the oxidation of n-heptane. For reasons of better visibility in this graph the scale of n-heptane concentrations below 600 K was chosen different to that displayed for higher temperatures. The addition of CO slightly inhibits the n-heptane oxidation at low temperatures due to the consumption of $\bullet OH$ radicals via reaction (102), whereas at higher temperatures, the CO addition accelerates the consumption of n-heptane and the NTC-amplitude decreases via the $\bullet OH$ producing reaction (103):

(102) $CO+\bullet OH \leftrightarrow CO_2+\bullet H$

(103) $CO+\bullet OOH \leftrightarrow CO_2+\bullet OH$

These observations are in agreement with the analysis performed by Subramanian et al. (2007) on the impact of CO on ignition delay times.

5.2.4. Concluding remark on the study on the formation of C_1-C_2 species

This study points out the influence of EGR compounds on the fuel oxidation in HCCI-engine applications. The experimental results and the developed kinetic model contribute to a deeper understanding of complex chemical impacts of EGR on the fuel oxidation control. Our study confirms the observations made by Dubreuil et al. (2005) and Moréac et al. (2006), who have emphasized the importance of the impact of NO on the oxidation of hydrocarbons. Generally, the presence of NO lowers the overall fuel reactivity at low temperatures and increases it at higher temperatures. Therefore, the content of recycled NO might impact strongly the ignition delay in an HCCI-engine. The addition of NO reduces considerably the formation of ethylene, ethane, CO and H_2 above 900 K and promotes those of CO and H_2O. Ethylene and methanol were experimentally and theoretically found to have small impact on the oxidation of n-heptane/toluene mixtures, even if a slight retarding effect on the fuel oxidation is observed at low temperatures. In contrast, a strong increase in the NTC-amplitude is predicted for the addition of formaldehyde. At high concentrations and at temperatures below 750 K, its presence can completely inhibit the fuel oxidation. It was further shown that CO influences the oxidation kinetics of hydrocarbons and that therefore the impact of CO on HCCI-engine control might not only be limited to a diluting effect. This is in agreement to the results shown in the first part of this chapter.

5.3   Ignition limits under postoxidation conditions

The analysis shown in the previous sections revealed that under postoxidation conditions, dilutant species may shorten ignition delays. The previous study was however performed in a temperature range between 600 and 950 K, corresponding to conditions generally observed in the exhaust line of internal combustion engines. At those temperatures and under high dilution ratios, the order of magnitude of ignition delays can exceed seconds. Such values are by far too long to explain, autoignition during secondary air injection. In a 4-stroke engine, exhaust gases do not remain in the exhaust pipe between the combustion chamber and the catalytic converter during more than 2 engine cycles. At low engine speed (1000 rpm), the time spent for two engine cycles is 0.12 seconds. For SAI ignition to occur before the catalytic converter, unburned hydrocarbons must then react within this period of time. Hence, we are interested in determining the thermochemical conditions (*pressure, temperature, equivalence ratio* and *dilution ratio*) under which postoxidation may occur (i.e. with a maximum ignition delay of 0.12 s).

We have computed ignition delays for the surrogate fuel defined in section 5.1.1 (page 86) with the reaction mechanism developed in chapter 4 under constant pressure conditions for different initial temperatures T, pressures P, equivalence ratios ϕ and the dilution ratios x_{res}. All tested reactive mixtures consisted of defined molar fractions of fuel, oxygen and N_2. Hence, the dilution ratio x_{res} corresponds here to the mole fraction of N_2.

$$[8] \quad x_{res} = x_{N_2}$$

This definition, however, does not correspond to the one used in the combustion model ECFM-3Z where the methodology developed in this work will be implemented as described in Chapter 7. The latter has the advantage of being close to the EGR ratios measured in an engine test bench. For the sake of simplicity, this quantity is called x_{EGR} and it is defined by:

$$[9] \quad x_{EGR} = x_{res} - \left(\frac{0.79}{0.21}\right) \cdot x_{O_2}$$

where x_{O2} is defined by the equivalence ratio ϕ, the fuel stoichiometric coefficient S.

$$[10] \qquad x_{O_2} = \frac{S}{\phi} \cdot \frac{1 - x_{res}}{1 + S/\phi}$$

Postoxidation computations are based on the exhaust gas composition issued from engine combustion. Hence, the fraction of N_2 issued from combustion air and the fraction of N_2 issued from recycled residual gases is unknown. Thus, a characterization of exhaust gases by a fraction of EGR x_{EGR} has no physical meaning. However, for IC-engine engineers, the fraction of EGR x_{EGR} is a commonly used parameter for the characterization of reactive systems in engines. For better understanding, Table 22 indicates the EGR mole fraction x_{EGR} corresponding to our definition of the fraction of residual gases x_{res}.

We have tested ignition delays over a wide range of conditions. The variations of pressure, temperature, equivalence ratio and dilution (x_{res}, x_{EGR}) are listed in Table 23.

Table 22 Mole fractions of EGR (x_{EGR}) as a function of the dilution ratio x_{res}.

x_{res} [mol/mol]	x_{EGR} [mol/mol]
0.89	0.52
0.91	0.60
0.93	0.69
0.95	0.78
0.97	0.86
0.99	0.96

Table 23 Range of tested thermodynamic conditions.

	Unit	Range
Pressure P	[bar]	0.5-3
Temperature T	[K]	800-1300
Equivalence ratio ϕ	[-]	0.5-6
Dilution ratio x_{res}	[-]	0.89-0.99
Ratio of EGR x_{EGR}	[-]	13%-60%

The fuel is the PRF/toluene mixture defined in Table 17. The ignition delays obtained for the variations of initial temperature and dilution ratio are visualized in Figure 5-15. The pressure was the atmospheric pressure and the equivalence ratio ϕ was kept equal to 1. The ignition limit of 0.12 s is shown by a plane. The projection in the T-x_{EGR} plane of Figure 5-15 marks the conditions in the T-x_{EGR} space under which postoxidation may occur in the exhaust line of an IC-engine. Ignition delays are shown in a logarithmic scale.

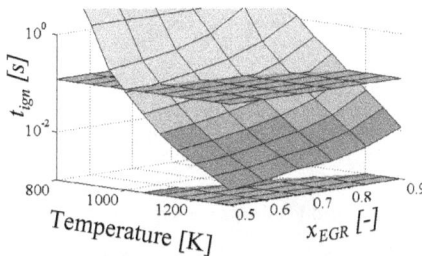

Figure 5-15 Ignition delay times as a function of initial temperature T and the EGR-fraction x_{EGR} for the oxidation of the n-heptane/iso-octane/toluene mixture, under stoichiometric conditions and at atmospheric pressure (constant pressure conditions).

As expected, temperature is the main parameter determining the ignition delay under such postoxidation conditions. It is observed that under any condition, temperatures for which ignition delays are short enough to ignite during 2 engine cycles, are higher than 1000 K. Dilution also impacts ignition delays and under highly diluted conditions, the temperature limit may shift up to 1200 K. In an IC-engine exhaust line, such high temperatures are not common, especially during cold start where the Secondary Air Injection technique can be used. Thus, a re-ignition of unburned hydrocarbons forced by SAI should be strongly limited when air is injected into the exhaust line only. However, SAI has indeed been observed. It should then occur when the engine exhaust valves open and an important 'back-flow' of exhaust gases and injected air into the cylinder is observed (Kleemann, 2006). Therefore, important fractions of the injected fresh air are mixed with hot gases inside the combustion chamber, where temperatures exceed those in the exhaust line. The ignition limits displayed in Figure 5-15 show that the 'back flow' of gases into the combustion chamber during exhaust gas opening must play a major role for postoxidation induced by SAI.

In an IC-engine exhaust line, pressure fluctuations of more than 1 bar may occur. We have thus also tested the influence of pressure on ignition delays. Figure 5-16 shows the computed ignition delays as function of dilution (x_{EGR}) and pressure. The initial temperature at which the analysis was performed was 1000 K and the equivalence ratio ϕ was equal to one. Ignition delays are shown in a logarithmic scale.

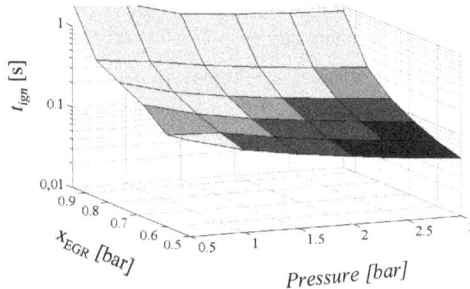

Figure 5-16 Ignition delays as a function of the EGR-fraction x_{EGR} and the initial pressure P for the oxidation of the n-heptane/iso-octane/toluene mixture, under stoichiometric conditions and at a temperature of 1000 K (constant pressure conditions).

Figure 5-16 reveals that ignition delays are impacted in the same order of magnitude by pressure changes and different degrees of dilution (keeping $T = 1000$ K and $\phi = 1$). The impact of the dilution ratio is more important at high degrees of dilution ($x_{EGR} > 0.75$). Generally, when pressure increases by a factor two, the ignition delay is reduced by almost the same factor under the tested conditions.

We have further studied the impact of the equivalence ratio on ignition delays. The degree of dilution x_{res} of the performed computations was 0.97 ($x_{EGR} \approx 0.8$) and the initial temperature 1000 K. The ignition delays obtained are presented in Figure 5-17.

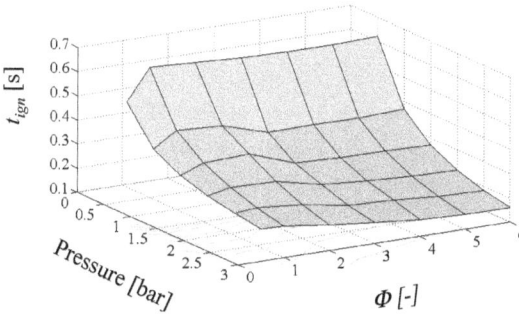

Figure 5-17 Ignition delay as a function of the initial pressure P and the equivalence ratio φ for the oxidation of the n-heptane/iso-octane/toluene mixture for an EGR-fraction $x_{EGR} \approx 0.8$ and at a temperature of 1000 K (constant pressure conditions).

The impact of changes in equivalence ratio φ is rather small compared to pressure variations. We thus conclude that ignition limits under postoxidation conditions are mainly determined by the temperature range at which the unburned hydrocarbons are mixed with air. In the case of SAI, the minimum temperature at which postoxidation may occur is greater than 1000 K. Such temperature limits may shift towards lower temperatures when the local pressure exceeds atmospheric pressure due to pressure fluctuations in the exhaust manifold. Also, the degree of dilution inside the combustion chamber may be much lower than that in the exhaust line due to fuel (unburned hydrocarbons) exiting crevices. Hence, during SAI, the 'back-flow' of exhaust gases and injected air into the combustion chamber strongly favours exhaust gas auto-ignition.

5.4   Conclusion on the performed numerical studies

The three studies performed allowed an estimation of the impact of the major compounds present in engine exhaust gases on hydrocarbon oxidation. The kinetic mechanism developed in chapter 4 was used to analyze chemical kinetics under thermochemical conditions typically found in the exhaust line of IC-engines.

The study of the influence of residual gases on hydrocarbon oxidation revealed that under postoxidation conditions CO_2 and H_2O show a kinetic impact when added into N_2. CO also shows a remarkable effect on hydrocarbon oxidation, when present in residual gases. In actual combustion models CO_2 and H_2O are considered as inert, not taking part in reaction kinetics. However, especially H_2O may impact reaction kinetics around 750 K as collision partner in 3rd-body reactions. At low temperatures the dissociation of H_2O_2 to •OH is rate governing and its rate constant depends strongly on the collision efficiencies of its collision partners. H_2O is an efficient collision partner and thus, impacts strongly H_2O_2-dissociation and global mixture reactivity at low temperatures. It was shown that collisions efficiencies play a major role for low temperature (T <900 K), but important uncertainties remain on such data. An improved comprehension of low temperature reaction kinetics demands further investigations on the low pressure limit rate constant of 3rd-body reactions.

Our study of the formation of C_1-C_2 species revealed that the predictivity of the standard EXGAS-version of the formation of ethylene is limited to temperatures greater than 800 K. At lower temperatures the production of ethylene is strongly overestimated. That inspired us to revise the secondary mechanism of EXGAS which allowed us to improve drastically the prediction of C_1-C_2 species formation at low temperatures. This is a significant optimization of our kinetic model, which enlarges its potential applications for CFD-combustion models. C_1-C_2 species are precursors of soot formation and a correct prediction of such species is fundamental for CFD-models describing the formation of soot during hydrocarbon oxidation. The quality of prediction of C_1-C_2 species formation shown in section 5.2 recommends the modified EXGAS-model as basis for CFD-models on soot formation. Nevertheless the proposed modifications in the secondary mechanisms were not integrated in the available EXGAS-version and for this reason all further calculations were performed with the mechanism proposed in chapter 4.

In the final section we tested the thermodynamic conditions under which postoxidation may potentially occur in an IC-engine exhaust line. We found that the temperature is the major parameter determining the probability of unburned hydrocarbon postoxidation. In the exhaust line commonly observed pressure fluctuations between 0.5 and 2 atm may change ignition delay times by more than a factor of 2. We concluded that postoxidation may occur at ignition temperatures greater ~1000 K, but pressure fluctuations may shift this temperature barrier towards lower temperatures.

Chapitre 6

Coupling of complex chemistry and CFD by FPI-Tabulation methodologies

The integration of the detailed chemistry into the CFD-code *IFPC3D* was performed by a tabulation methodology. Our approach is a FPI-like tabulation of chemical kinetic variables under postoxidation conditions which is based on the detailed kinetic mechanism developed in this work. We describe here the principles and methods used for coupling complex chemistry with CFD. We will first introduce the principles of tabulations of complex chemistry and present the methodology of FPI tabulation. The issues related to tabulating complex chemistry under postoxidation conditions will then be emphasized. Limits of the tabulation methodology will be outlined and the consequences for the postoxidation tabulation will be presented.

6.1   Fundamentals of tabulating complex chemistry by FPI methods

In this section, the methodology of FPI tabulation is introduced and the basics of FPI tabulation are presented.

6.1.1. The CFD code IFP-C3D and the combustion model ECFM3Z

The CFD-Code used in this work is IFP-C3D which was developed at IFP. The code is a Reynolds Averaged Navier-Stokes (RANS) transient, three-dimensional, multiphase and multi-component code for the analysis of chemically reacting flows. It uses an Arbitrary Lagrangian Eulerian (ALE) methodology on a staggered grid and discretizes space using the finite-volume technique. Its field of validity ranges from low speed to supersonic flows for both, laminar and turbulent regimes. An arbitrary number of species and chemical reactions is allowed. Although specifically designed for performing internal combustion engine calculations, the modularity of the code allows easy modifications for solving a variety of hydrodynamics problems involving chemical reactions.

For turbulent combustion simulations, the ECFM3Z model (Extended Coherent Flame Model – 3 Zones) of Colin et al. (2003) is used. ECFM3Z is based on a flame surface density equation which takes into account the wrinkling of the flame front surface by turbulent eddies and a conditioning and averaging technique allowing precise reconstruction of local properties in fresh and burned gases. This model has been used successfully for simulating both gasoline and diesel engine applications (Duclos et al. 1996; Duclos and Zolver, 1998; Lafossas et al. 2002; Henriot et al. 2003; Kleemann et al. 2003).

The ECFM3Z-model is structured in three mixing zones:

- A pure fuel zone.

- A pure air plus possible residual gases zone.

- A mixed air and fuel zone.

These three mixing zones are separated by the premixed flame into regions of burned (*b*) gases and unburned (*u*) gases. Figure 6-1 shows schematically the mixing zones of ECFM3Z and their separation in unburned and burned gases.

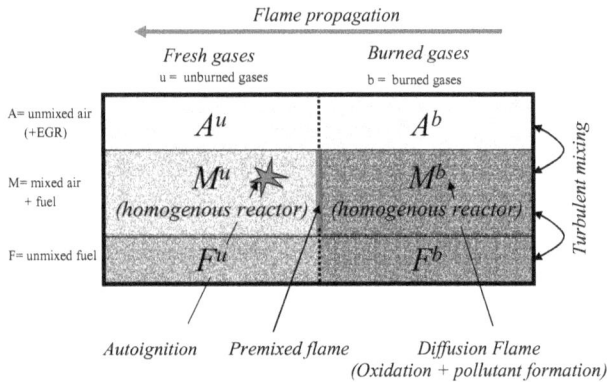

Figure 6-1   Scheme of the mixing zones in the ECFM3Z-model (Colin, 2004).

Auto-ignition is modelled following Colin et al. (2004) and the premixed turbulent flame description is given by the ECFM3Z-model. In the case of postoxidation, all species belong to the burned gases zone (region M^b shown in Figure 6-1) and are also supposed to be perfectly mixed. In that zone, chemical reactions are modelled by the Flame Prolongation of ILDM (FPI) methodology.

6.1.2. The Flame Prolongation of ILDM (FPI) model

In 1992, Maas and Pope (1992) have demonstrated the existence of attractive trajectories in a chemical reaction process. Whatever the initialization, a chemical system joins within more or less time an attractive trajectory towards reaction equilibrium in the composition space. Thus, Maas and Pope have proposed the method of *Intrinsic Low Dimensional Manifolds* (ILDM) which simplifies chemical kinetics by computing such trajectories (manifolds). All ILDM are stored in a look-up table that can be read by a CFD-code with negligible costs in CPU (Central Processor Unit) time compared to a direct calculation of complex chemistry. Figure 6-2 presents such an ILDM in the composition space y_1, y_2.

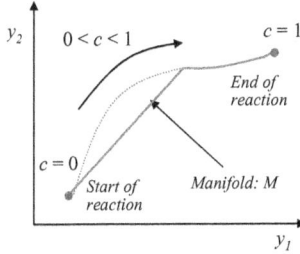

Figure 6-2 Intrinsic Low Dimensional Manifold M in the composition space y_1, y_2 as proposed by Maas and Pope (1992).

The state of reaction is defined by the starting point of its trajectory and by a single progress variable c varying from 0 to 1 towards reaction equilibrium. The reaction starts at point '$c = 0$' and ends at point '$c = 1$' in the composition space. The thin dotted line represents the trajectory (manifold) of the real reacting mixture, the thick full line the manifold reconstructed by the ILDM method. ILDM uses chemical characteristic time scale analysis to construct low-dimensional surfaces (manifolds) in the composition space for given thermodynamic conditions (pressure, temperature, fuel/air equivalence ratio ....).

Gicquel et al. (2000) have observed that the ILDM-method of Maas and Pope is insufficient for low temperature applications. They have proposed a Flame Prolongation of ILDM (FPI) by 1D-flame calculations and have shown that the obtained FPI-manifolds represent well the low temperature regime. Figure 6-3 compares ILDM and FPI in the composition space.

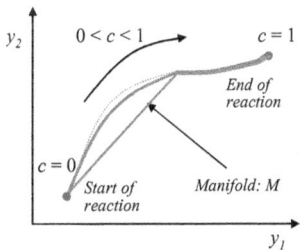

Figure 6-3 Comparison of ILDM and Flame Prolongation of ILDM (FPI) in the composition space y_1-y_2. The thin dotted line represents observed chemistry, full grey line ILDM proposed by Maas and Pope (1992), the thick red line FPI proposed by Gicquel et al. (2000).

Like in Figure 6-2, the thin dotted line represents the trajectory (manifold) of the real reacting mixture, the thick full line the manifold reconstructed by the ILDM method proposed by Maas and Pope (1992). Additionally, the thick red line illustrates a trajectory as it is obtained by the FPI-method proposed by Gicquel et al. (2000). Embouazza (2005) replaced FPI premixed flame calculations by auto-ignition calculations and obtained good predictions of auto-ignition. Pera et al. (2009) were the first to apply the FPI model to engine combustion and used it successfully for the prediction of engine pollutants.

6.1.3. Construction of a FPI look-up table

From the FPI-assumption of one unique manifold for a reactive system during reaction, it follows that the thermodynamic state of composition which results from a given initial condition is defined by one progress variable c. The state of a reactive system is stringently defined by the progress variable c and its initial thermodynamic condition at the beginning of the reaction. Such a reactive system may be represented by a closed homogenous reactor. Each dimension in the thermodynamic space corresponds to one dimension in the look-up table. The look-up table must cover the complete thermodynamic space which potentially might be achieved during a CFD-calculation in the studied system. In an internal combustion engine, the thermodynamic space of a reactive gas-mixture is defined by the dimensions:

- Temperature (T_0)
- Pressure (P_0)
- Fuel/air equivalence ratio (ϕ_0)
- Dilution mass fraction (y_{res})

For each potential initial state, one closed homogenous reactor is computed. As the solutions used in the CFD-model will be linearly interpolated from the stored solutions in the FPI look-up table, each dimension of the table must be discretized with sufficient accuracy in order to minimize interpolation errors. Figure 6-4 illustrates the construction of the look-up table as well as the information stored in it.

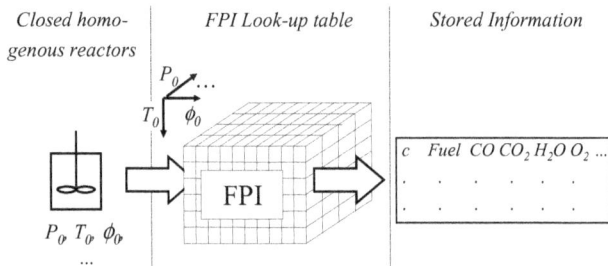

Figure 6-4   Principle structure of a FPI look-up table.

The thermodynamic space is discretized along its dimensions (P_0, T_0, ϕ_0...) and for each table point (corresponding to one thermodynamic condition at reaction start, $c = 0$), the species concentrations (manifolds) are stored along c, discretized between 0 and 1. The tabulated thermodynamic dimensions (P_0, T_0, ϕ_0...) correspond to the thermodynamic conditions at reaction start ($c = 0$) and allow the localisation of the corresponding trajectory in the FPI-table. For a given computational cell in the CFD code corresponding to a set of thermodynamic conditions, the initial state of the reaction has to be tracked in order to retrieve its corresponding values from the *a priori* table. Initial conditions, P_0, T_0, ϕ_0... are then transported as tracer-characteristics in the fresh gases. Tracer characteristics P_0, T_0, ϕ_0 as well as tracer species (y^T) are virtual parameters without any physical interaction. They are transported as virtual non-reacting flow properties, thus allowing the reconstruction of the initial state of reaction at $c = 0$ at any state of reaction. The reconstruction of initial conditions at $c = 0$ by tracer species and tracer characteristics is fundamental for the location of the right manifold in the FPI-table.

For reasons of CPU-time cost and memory limitations, the number of transported species in the CFD code is limited and so is the number of species stored in the look-up table. The calculations for building the table are performed in a closed homogeneous reactor at constant pressure. This is reasonable, as in ECFM3Z postoxidation takes places in the burned gases zone under premixed conditions (M^b in Figure 6-1) corresponding to reaction conditions of a closed homogeneous reactor. Our FPI-tabulation strategy consists in tabulating species mass fractions instead of reaction rates as proposed originally by Gicquel et al (2000). Such a species concentrations tabulation FPI model was developed at IFP by Pera et al. (2009) and was successfully applied to engine applications.

The kinetic mechanism on which our FPI-tabulation is based consists of around 540 species and 3000 reactions. This is by far too exhaustive to be handled directly by a CFD-code where the number of transported and tabulated species is limited. Hence, main species representative of potential states of reaction ($0 < c < 1$) have to be determined in order to reduce the size of the handled system. The choice of the main species is crucial as according to the main FPI assumption, their characteristics should be representative of the complete reactive system. By considering aspects of mass and energy correlations, Pera et al. (2009) identified for the combustion of n-heptane 9 major species of which the mass fraction are transported. These are the following:

- y_{C7H16},	- y_{O2}	- y_{CO}
- y_{CO2}	- y_{H2O},	- y_H
- y_{N2}	- y_{H2}	- y_{C7H14}.

This reduction towards a few main species unavoidably leads to loss of mass – and energy – conservation. To overcome this difficulty, atomic conservation equations are written. For hydrocarbon combustion, H, C and O-atoms balance equations are considered. The balance equations are closed by non tabulated reconstructed species. These are artificial species and represent the mass of all non tabulated quantities. As a consequence, their mass fraction differs significantly from that given for the same species by the complex chemistry. Pera et al. (2009) have chosen C_7H_{14}, n-heptene, for closing the atomic C-balance equation, H_2 for closing the atomic H-balance equation and O_2 for closing the O-balance equation. Therefore C_7H_{14}, H_2 and O_2 are deduced from the following atomic balance equations:

$$[11] \quad y_{O_2} = \frac{M_{O_2}}{2}\left[2 \cdot \frac{y^T_{O_2}}{M_{O_2}} - \left(\frac{y_{CO}}{M_{CO}} + 2 \cdot \frac{y_{CO_2}}{M_{CO_2}} + \frac{y_{H_2O}}{M_{H_2O}}\right)\right]$$

$$[12] \quad y_{C_xH_y} = \frac{M_{C_7H_{14}}}{7}\left[x \cdot \frac{y^T_{C_xH_y}}{M_{C_xH_y}} - \left(x \cdot \frac{y_{C_xH_y}}{M_{C_xH_y}} + \frac{y_{CO}}{M_{CO}} + \frac{y_{CO_2}}{M_{CO_2}}\right)\right]$$

$$[13] \quad y_{H_2} = \frac{M_{H_2}}{2}\left[y \cdot \frac{y^T_{C_xH_y}}{M_{C_xH_y}} - \left(y \cdot \frac{y_{C_xH_y}}{M_{C_xH_y}} + 7 \cdot \frac{y_{C_7H_{14}}}{M_{C_7H_{14}}} + 2 \cdot \frac{y_{H_2O}}{M_{H_2O}} + \frac{y_H}{M_H}\right)\right]$$

where y_i is the mass fraction of species i, M_i its molar weight, and y_i^T is the concentration of the tracer species i deduced from the ECFM3Z model. By definition, the tracer of a species i is convected and diffused exactly like its corresponding real species. However, unlike the real species i, the pseudo-species is neither consumed during combustion nor accounted for thermodynamic balances. Hence, its transport equation is the same as the one for i, but neglecting the reaction term (Colin et al. 2004). The reconstructed species have to be chosen carefully as their thermal properties contribute to the total system internal energy which is defined as follows:

$$[14] \quad e = e_s + \sum_{i=1}^{N} \Delta h_i^f y_i = const.$$

where h_i^f is the standard formation enthalpy and y_i the mol fraction of species i. e_s denotes for the sensitive mixture energy and is defined as:

$$[15] \quad e_s = \int_{T_0}^{T} C_v dT - \frac{RT_0}{W}$$

with C_v the heat capacity at constant volume, T_0 the reference temperature R the perfect gas constant and W the mean molecular weight. The standard enthalpy of formation of a compound is defined as the change of enthalpy that accompanies the formation of 1 mole of a substance in its standard state from its constituent elements in their standard states (the most stable form of the element at 1 bar of pressure and the specified temperature, usually 298.15 K).

6.1.4. Definition of the progress variable c and reaction progress

It was stated that the FPI-methodology implies the assumption that a state of reaction is defined by its thermodynamic conditions at reaction start and by a progress variable c determining the reaction progress. In consequence, c must fulfil stringently the following criteria:

- c must be monotonic in time.

- Under any condition, c must be representative for the reactive system.

Previous studies have suggested different definitions of the progress variable c. Fiorina et al. (2003), Colin et al. (2004, 2005) and Ribert et al. (2006) have proposed a definition of c based on a linear combination of CO and CO_2 mass fractions (y_{CO} and y_{CO2}). For a better reproduction of cool-flame behaviour, Embouazza et al. (2003) have added the fuel mass fraction to the definition of c, while for the same reason, Mauviot et al. (2006) have proposed a combination with the oxygen mass fraction rather than fuel (see appendix B.1). The most current definition of the progress variable c is that described by equations [16] and [17].

$$[16] \quad y_c = y_{CO} + y_{CO_2}$$

$$[17] \quad c = \frac{y_c - y_c^0}{y_c^{eq} - y_c^0}$$

where y_c^0 and y_c^{eq} are the initial and the equilibrium mass fractions of the y_c advancement. However, in the case of postoxidation, CO and CO_2 are already present in the reacting mixture at reaction start and non zero. Therefore, when CO and CO_2 are non zero at reaction start, the CO and CO_2 produced from fuel consumption must be distinguished from the CO and CO_2 present prior to combustion (residual species y_{res}):

$$[18] \quad y_i^{tot} = y_i + y_i^{res}$$

ECFM3Z treats the residual species mass fractions y_i^{res} by specific transport equations (Colin et al. 2004) and the mass fractions y_i^{res} of the residual species are subtracted from the total mass fraction y_i^{tot} before the progress variable is calculated by equations [16] and [17]. As described before, we have chosen the method of tabulating species mass fractions rather

---

than reaction rates. Therefore, these have to be determined as they are present in the species transport equations.

Following Pera et al. (2009), species source terms are reconstructed from tabulated species mass fractions by the approximation:

$$[19] \qquad \omega_i = \frac{\delta y_i}{\delta t} = \frac{y_i^{FPI}(c(t+\tau)) - y_i(t)}{\tau}$$

where y_i denotes the species mass fraction and y_i^{FPI} is the FPI tabulated species. t is the current time and τ is a characteristic time scale allowing to relax towards the tabulated chemical manifold path. τ has to be chosen greater than the current time step dt, but very small compared to the characteristic combustion time. Thus, τ is limited by:

$$[20] \qquad dt < \tau < \frac{1-c}{\omega_c}$$

where ω_c is the time derivative of the progress variable c. Equation [19] allows FPI table trajectories to relax towards tabulated manifolds within the characteristic relaxation time τ. The term $c(t+\tau)$ in equation [19] is the progress variable at the time $t+\tau$. It is approximated by a linear interpolation:

$$[21] \qquad c(t+\tau) = c(t) + \tau \cdot \omega_c^{FPI}$$

where ω_c^{FPI} is the progress variable reaction rate, which is calculated by the gradient between two consecutive points j and $j-1$ in the FPI table.

$$[22] \qquad \omega_c^{FPI} = \frac{c_j^{FPI} - c_{j-1}^{FPI}}{t_j - t_{j-1}}; c \in |c_{j-1}; c_j|; j = 2...N$$

with t_j, the tabulated reaction time and c_j^{FPI} the corresponding tabulated reaction advancement. The mass fractions at $t+dt$ are then obtained by:

$$[23] \qquad y_i(t+dt) = y_i(t) + \omega_i \cdot dt = y_i(t) + \frac{y_i(t+\tau) - y_i(t)}{\tau} \cdot dt$$

The FPI-progress of mass fractions in time is explained in detail in appendix B.2.

After updating the tabulated species mass fractions y_i (for example: y_{C7H16}, y_{CO}, y_{CO2}, y_{H2O}, y_{NO}, y_H) at time $t+dt$, the reconstructed species mass fractions y_i^{res} (for example: y_{C7H14}, y_{H2}, y_{O2}) are calculated and the mass (equations [11]-[12]) and energy balances (equation [13]) are closed.

6.1.5. Table implementation into the CFD-Code

The CFD-code calls the FPI-table for every computed time step. Figure 6-5 shows a principle sketch of the coupling interactions between a FPI-table and the CFD-code IFP3D.

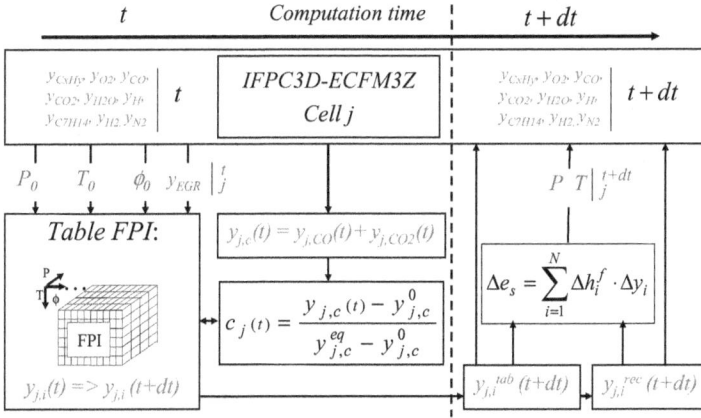

Figure 6-5 Scheme of the coupling between a FPI-table and the CFD-code IFP3CD.

The i species mass fractions $y_i(t)$ of the tabulated species (y_{C7H16}, y_{CO}, y_{CO2}, y_{H2O}, y_{NO}, y_H) plus the three species mass fractions $y_i^{rec}(t)$ of the reconstructed species (y_{C7H14}, y_{H2}, y_{O2}) are transported by the CFD-code. For each computational cell j, the progress variable $c(t)$ is computed at time t by equations [16] and [17] based on the transported mass fractions y of the species CO and CO_2. The thermodynamic dimensions (P_0, T_0, ϕ_0, y_{EGR}) are computed for a virtual non-reactive mixture by the tracer species $y_i^T(t)$.

Hence the table dimensions P_0, T_0, ϕ_0 ... allow the localisation of the corresponding tabulated trajectory defined for $c = 0$ to $c = 1$. P_0, T_0, ϕ_0 etc. may vary locally (for different computational cells j) and also in time t due to species transport. Depending on $c(t)$, the mass fractions $y_i(t)$ of each tabulated species are advanced according to equations [19]-[23] and the tabulated mass fractions $y_i^{tab}(t+dt)$ are computed. Mass and element balances are closed by recomputation of the reconstructed species mass fractions $y_i^{rec}(t+dt)$ by equations [11]-[13]. The updated mass fractions $y_i(t+dt)$ are given to the CFD code, the thermodynamic conditions ($P(t+dt)$, $T(t+dt)$) are recalculated as a function of the heat release during dt obtained via equation [14].

6.2   Tabulation of complex chemistry under postoxidation conditions

The FPI-methodology described in the last section was adapted and applied to the specific demands of postoxidation computations. The following section compares the needs for FPI-tabulation for IC-engine computations with those for the case of postoxidation applications in an IC-engine exhaust line. Based on these needs, our FPI-tabulation strategy was developed and coupled to IFP-C3D. The adapted FPI-methodology is presented here.

6.2.1. Identification of determinant thermodynamic conditions

A major challenge for the generation and application of FPI-look up tables is the identification of the initial thermodynamic conditions $(P_0(t),\ T_0(t),\ \phi_0(t)\ldots)$ at which a mixture starts to react. As the thermodynamic conditions in the exhaust line differ strongly from those found during IC-engine combustion inside a combustion chamber, the architecture of a FPI look-up table designed for postoxidation conditions differs strongly from that of tables designed for engine computations. A comparison of table architectures for both kinds of FPI-applications is shown in Table 24.

Table 24 Comparison of FPI-table architectures designed for IC-engine applications and postoxidation applications.

		FPI-Engine		FPI-Postoxidation	
		Min.	Max.	Min.	Max.
Thermodyn. Conditions	P_0 [atm]	1	80	0.5	2
	T_0 [K]	500	1400	600	1200
	ϕ_0 [-]	0.4	3	0	∞
	y_{res} [-]	0	0.7	0.7	1.0
Table dimension	[-]	$P_0,\ T_0,\ \phi_0,\ y_0^{res}$		$P_0,\ T_0,\ \phi_0,\ CO_0,\ CO_{2\_0},\ H_2O_0,$ $NO_0,\ SAI_0$	
Table size	[-]	$\sim 10^5$		$> 10^6$ points	
Initial Condition.	[-]	well defined		based on accumulated errors during engine calculations	

In case of IC-engine computations inside the combustion chamber, the look-up table conditions are rather well defined. Pressure and temperature are well known from CFD-calculations or defined initial conditions. The mixture composition is given by the fresh gas composition containing fuel and combustion air. The degree of dilution is determined by the fraction of recirculated and trapped burned gases inside the combustion chamber (EGR). Generally, EGR compounds are considered as non-reactive and the effect of EGR residual gases on the combustion process is taken into account only by their different heat capacity C_p compared to N_2.

For a FPI-tabulation of an IC-engine combustion the determining table dimensions are:

Thermodynamic:

1. Pressure P_0
2. Temperature T_0

Composition:

3. Equivalence ration ϕ_0
4. Dilution mass fraction y_0^{res}

However, the ranges of pressure and temperature which have to be covered by such a FPI-look-up table are huge. Inside the combustion chamber, pressures may vary between 1 and 80 atm and temperatures at which mixtures may start to react range between 500 K and 1200 K. These conditions exceed the range of conditions under which detailed reaction mechanisms have been validated. In the case of postoxidation applications, the range of thermodynamic conditions is much smaller. In the exhaust line, the pressure varies around the atmospheric pressure and potential temperatures may range from 600 to 1200 K. These conditions are inside the range of conditions for which detailed reaction mechanisms have been validated. Nevertheless, the composition of the reactive mixture is much more complex than that of the fresh gases inside the combustion chamber. In contrast to IC-engine combustion, during postoxidation, the reactive mixture is not only composed of fuel and combustion air, but also of all other burned gas components produced during the IC-engine combustion period before. That means that the burned gas main components CO, CO_2, H_2O, NO have to be considered. Our kinetic studies on the impact of dilutant gases (chapter 5) have revealed the chemical impact of these species on hydrocarbon oxidation. It was shown that reducing their impact on reaction kinetics to thermal effects only, oversimplifies the problem under postoxidation conditions. Thus, the impact of these compounds should be tabulated as well. In the postoxidation case, the determining table dimensions are:

Thermodynamic:	1. Pressure (P_0)
	2. Temperature (T_0)
Composition :	3. Fuel$_0$ mass fraction (y_{CxHy_0})
	4. O_{2_0} mass fraction (y_{O2_0})
	5. CO_0 mass fraction (y_{CO_0})
	6. CO_{2_0} mass fraction (y_{CO2_0})
	7. H_2O_0 mass fraction (y_{H2O_0})
	8. NO_0 mass fraction (y_{NO_0})

In the case of Secondary Air Injection (SAI), a further dimension of the SAI-fraction has to be considered:

9. SAI – fraction (α_0)

with α_0 defined as :

$$[24] \qquad \alpha = y_{N_2}^{SAI} \cdot \left(1 + \frac{0.23}{0.77}\right)$$

However, each additionally tabulated mixture compound enlarges the table architecture by one dimension and thus the number of stored table points. The order of magnitude of stored table points increases from around 10^4 in the case of IC-engine tables up to 10^6 and more in the postoxidation case. The principle of the FPI-methodology is based on linear interpolations between the tabulated manifolds and reaction states. An increase of table dimensions by one results in doubling the number of interpolation steps. Figure 6-6 displays the order of magnitude of interpolation CPU-time costs as a function of the number of table dimensions. The computation costs are normalized to the calculation effort necessary for a mono-dimensional linear interpolation.

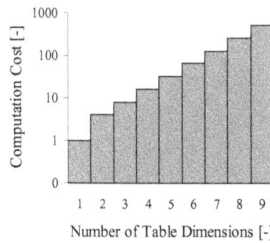

Figure 6-6   Interpolation calculation costs in function of the table dimensions.

It is obvious that there is a limitation to the maximum number of table dimensions due to the correlated quadratic increase of computational interpolation effort.

6.2.2. Assumptions resulting from FPI engine computations

The initial gas composition differs strongly under postoxidation conditions from the initial gas composition inside an IC-engine. While in the case of IC-engine computations, the mixture composition is well defined by the initial engine conditions, in the case of postoxidation, the mixture composition is only known from earlier results of IC-engine computations. Consequently, the initial conditions of postoxidation calculations are based on the uncertainties and the accumulated errors of the previous IC-engine simulation. The challenge of postoxidation FPI-calculations therefore consists in a reduction of table dimensions towards a reasonable table size as well as in handling uncertainties of previous IC-engine simulations. Different types of uncertainties are accumulated during FPI-based engine computations. Since the postoxidation solutions are based on the ones obtained from engine outputs, an estimation of accumulated uncertainties during the engine-combustion phase is fundamental for the comprehension of postoxidation computations.

6.2.2.1. Uncertainties in the exhaust gas-composition

During engine computations based on the FPI-methodology, two kinds of errors in fuel gas composition are accumulated:

- *Uncertainties in the composition of partially oxidized compounds*

We have explained in section 6.1.2 the principle of species reconstruction by closure of element balances. This method implies that for each element (C, O, H), all partially oxidized, non tabulated species are gathered in one unique reconstructed species. However, the composition of partially oxidized compounds during hydrocarbon oxidation depends strongly on the combustion conditions. Under very rich conditions, due to the lack of oxygen, the fuel is only partially oxidized leading to different intermediate species. These compounds are large molecules with a carbon number close to that of the initial fuel. On the other hand, under slightly rich conditions, the fuel is almost completely oxidized and partially oxidized compounds at the end of combustion consist mainly of small hydrocarbons. Figure 6-7 illustrates that by showing the computed composition of partially oxidized compounds as a function of equivalence ratio ϕ. Results were obtained in a closed homogeneous reactor at atmospheric pressure. The simulations were based on the developed mechanism for the oxidation of PRF-Toluene mixtures. The initial fuel was the surrogate gasoline chosen for our study in chapter 5.

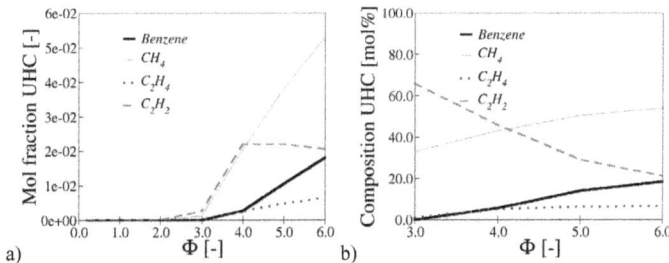

Figure 6-7 Composition of partially oxidized compounds in function of the equivalence ratio ϕ. (P = 1 bar, T = 800 K): a) Mol fraction of benzene (—), CH_4 (—), C_2H_4 (---) and C_2H_2 (---) in the burned gases, b) Relative composition of benzene (—), CH_4 (—), C_2H_4 (---) and C_2H_2 (---)) in the burned gases.

By accumulating the partially oxidized compounds within the chosen reconstructed species C_7H_{14}, H_2, O, the information about the composition of the partially oxidized species gets lost. One should note that due to the lack of oxygen under rich conditions the mass fractions of C_2H_2 exceed those of C_2H_4.

- *Uncertainties in the fuel composition*

We mentioned that for the CFD-computations, a fuel surrogate composed of 13.7 mol% n-heptane,   42.9 mol% iso-octane   and   43.4 mol% toluene   was   chosen (see chapter 3.1). In the CFD code *IFPC3D*, during the combustion phase, multi-component fuels are represented by one unique fuel species with an equivalent H/C-ratio. In the case of our three species surrogate fuel, the mass fraction of the fuel-species is calculated by:

$$[25] \quad y_{C_{7.55}H_{13}} = 0.137 \cdot y_{C_7H_{16}} + 0.429 \cdot y_{C_8H_{18}} + 0.434 \cdot y_{C_7H_8}$$

Consequently, the $C_7H_{16}/C_8H_{18}/C_7H_8$-ratio is kept constant during the full combustion process. This assumption might be critical, as linear and branched alkanes oxidize faster than aromatic compounds and therefore, after a rich combustion, the unburned HC are mainly composed of aromatics. Figure 6-8 shows the evolution of fuel composition computed during the oxidation of a mixture containing 13.7 mol% n-heptane, 42.9 mol% iso-octane and 43.4 mol% toluene. The equivalence ratio ϕ at which the fuel composition was tested was 6. Concentrations profiles are plotted against the non-dimensional time τ normalized defined in equation [6] (page 88).

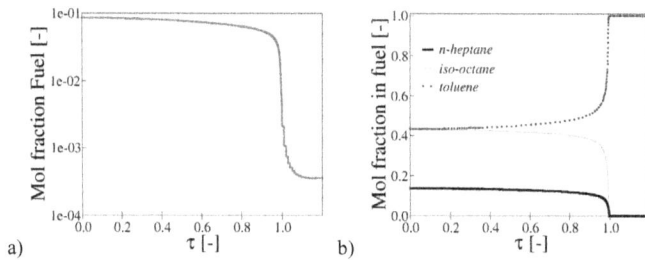

Figure 6-8 Evolution of fuel concentration and composition during the oxidation of 13.7 mol% n-heptane, 42.9 mol% iso-octane and 43.4 mol% toluene at atmospheric pressure, a temperature of 800 K and at an equivalence ratio $\phi = 6$. a) Evolution of the fuel mol fraction during combustion, b) evolution of the fuel composition during combustion.

It is obvious that the alkanes, n-heptane and iso-octane oxidize faster than the aromatic compound, toluene. A constant growth of the toluene distribution in the unburned fuel at the early stages of oxidation is the result of the complete oxidation of the n-heptane and iso-octane, so that at reaction equilibrium, the non-oxidized fuel consists of nearly 100% toluene.

6.2.2.2.   Uncertainties from neglecting fuel evaporation from walls

The composition of the non-oxidized fuel contained in the exhaust gases is crucial for a correct reproduction of postoxidation kinetics. Figure 6-9 shows the calculated temperature evolutions for the reaction of exhaust gases issuing from a rich combustion and then mixed with air in such a way that the equivalence ratio is stoichiometric. Exhaust gases are composed of about 500 different species included in the reaction mechanism. The reactivity of such a mixture is compared to the reactivity of a mixture reconstructed from the FPI-algorithms described in section 6.1 consisting only of n-heptane, iso-octane, toluene, CO, CO_2, H_2O, N_2, and O_2.

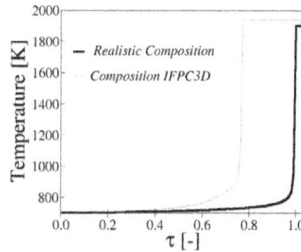

Figure 6-9   Temperature evolution of the stoichiometric reaction of exhaust gases mixed with air at an equivalence ratio 1 and a temperature of 750 K. The black line (—) corresponds to the temperature evolution obtained with the real composition (~500 species), the green line (—) corresponds to the temperature evolution obtained for the mixture reconstructed from the FPI-algorithms.

Comparison of the temperature evolutions of the reconstructed gas composition and the mixture composed of ~500 species shows a reduction of ignition delays of about 30% in the first case. This is mainly due to the presence of alkanes (n-heptane and iso-octane) in the reconstructed mixture, while the mixture composed of ~500 species mainly contains a larger amount of the less reactive aromatic compounds (compare Figure 6-8). The fuel represented by a surrogate containing n-heptane and iso-octane is more reactive.

Since our model is based on an a priori tabulation procedure, limited by the number of table input variables, it is not possible to include the total burned gas composition among those variables. An assumption has therefore to be made. Despite the observation that non-oxidized fuel of an initial alkane/aromatic surrogate consists at the end of combustion of mainly aromatic compounds, we still choose a surrogate composed of 13.7 mol% n-heptane, 42.9 mol% iso-octane and 43.4 mol% toluene (initial engine-fuel) to represent the non-oxidized fuel. This choice is justified as during spark-ignition IC-engine combustion, most unburned hydrocarbons (UHC) are desorbed from the lubricant film on the piston walls and combustion chamber crevices (engine head, piston segments…). During the expansion stroke, desorbed fuel evaporates into the burned gases before postoxidation occurs. Postoxidation fuel is thus, mostly composed of non-oxidized initial fuel making our assumption reasonable.

6.2.2.3. Uncertainties resulting from mass/mole fraction conversions

The CHEMKIN software used for a priori table construction computes complex chemistry in mole-fractions. The CFD code IFP-C3D uses on the other hand transported mass fractions. Therefore, when coupling detailed chemistry to the combustion model ECFM3Z, special care has to be taken on the conversion of species mass to species mole fractions. The conversion from mass to mole fractions is given by:

$$[26] \qquad x_i = \frac{\dfrac{y_i}{MW_i}}{\sum \dfrac{y_i}{MW_i}}$$

The correct conversion from mass to mole fractions thus implies the information on the total number of moles present in the mixture. When partially oxidized species are accumulated in reconstructed species, this information is lost and as a result, conversion of mass fractions in IFP-C3D to mole fractions will be different from the molar composition computed by the detailed chemistry, including all species in the kinetic mechanism.

$$[27] \qquad \sum \frac{y_i}{MW_i}\bigg|_{reconstructed} \neq \sum \frac{y_i}{MW_i}\bigg|_{chemistry}$$

However, due to the transport of mass fractions in IFP-C3D the table generation has to be performed in mass fractions as well. Thus, the conversion from mass to mole fractions must be stringently respected during table generation.

6.3 Concluding remarks

This chapter has introduced the basics of the FPI-tabulation methodology of detailed kinetic reaction mechanisms. The principles of ILDM and reaction progress were presented and the methods of storing chemical reaction data in a FPI-table have been explained. The principles of coupling such a database to a CFD code were shown. Attention was given to different issues leading to difficulties in tabulating postoxidation chemistry. We have therefore compared the cases of FPI-tabulations dedicated to IC-engine combustion conditions against FPI-tabulations applied to postoxidation conditions. The fundamental difference between the two is the definition and accuracy of initial computation conditions. In the case of IC-engine combustion, only a few thermochemical characteristics have to be tabulated such as, Pressure (P_0), Temperature (T_0), equivalence ratio (ϕ_0) and the dilutant mass fraction (y_{res}). When postoxidation kinetics are tabulated, the number of table dimensions and consequently the table size increases drastically as the initial composition of burned gases becomes also an important input (initial fuel mass fraction y_{CxHy_0}, initial O_2 mass fraction y_{O2_0}, initial CO mass fraction y_{CO_0}, initial CO_2 mass fraction y_{CO2_0}, initial H_2O mass fraction y_{H2O_0}, initial NO_0 mass fraction y_{NO_0}, ...). The importance of these parameters was studied and shown in chapter 5. However, table size and dimensions are limited and not all potentially relevant table characteristics may be tabulated. Another major difference is the necessary accuracy of initial conditions when dealing with postoxidation. While in the case of IC-engine tabulations initial conditions are well known, initial conditions for postoxidation depend on uncertainties issued from the previous IC-engine computations. These uncertainties result from accumulated errors during IC-engine computations. Major uncertainties from IC-engine combustion are errors in the fuel composition, errors in the composition of partially oxidized compounds, and uncertainties of fuel mass fractions resulting from evaporated fuel from walls and crevices. Further problems concern the conversions of mass to mole fractions. At the end of IC-engine computations, which is the start of each postoxidation simulation, these errors reach a maximum and present a major uncertainty for the initial conditions of postoxidation computations. A FPI-table predictive for postoxidation conditions has to deal with the challenges described above. Initial postoxidation conditions have to be reconstructed correctly from the information obtained from previous IC-engine computations without exceeding the limits of table size and number of table dimensions.

Chapitre 7

Construction of the FPI-Table for postoxidation applications

In the previous chapter, we have presented the FPI method and discussed the different assumptions necessary to apply such methodology to postoxidation conditions. Here, we describe the adapted FPI tabulation model. The first section deals with the choice of table dimensions and the associated assumptions. We present the chosen definition of the progress variable c and the choice of reconstructed species. Alternative definitions of the progress variable c and different types of reconstructed species are discussed and explained. It is shown that the FPI postoxidation model requires a re-initialisation of the species composition field resulting from the previous IC-engine computation. The re-initialisation procedure is explained. Model restrictions will be discussed and we show validations of the FPI tabulated chemistry against detailed chemistry computations.

7.1   Physical relations between possible table inputs

In section 6.2.1, we have presented the limitations associated to the look-up table size and the resulting restrictions in the number of table dimensions. It was outlined that for reasons of computational time and memory costs, the table dimensions should be minimized. The order of magnitude of stored table points easily may exceed 1×10^6 and the computational effort necessary for table interpolations increases drastically (see Figure 6-6). For such reason, a reduction of the number of table dimensions is unavoidable. A sensitivity analysis of reaction kinetics to the thermochemical characteristics, temperature, pressure and various species concentrations has been performed and is presented in appendix B.3. It has shown that none of these parameters should be neglected for a model describing postoxidation conditions. However, limitations of computational memory impose that a maximum number of table dimensions of seven should not be exceeded. Our strategy is thus to check for physical relations between possible table inputs and to couple them when possible. This approach may allow a reduction of table dimensions avoiding the loss of important physical information. The following sections analyse the possible table dimensions and the physical relations between them.

7.1.1. Equivalence ratio

When IC-engine conditions are tabulated, fuel and oxygen are linked by the equivalence ratio (Pera et al. 2009). This is reasonable because the ratio of fuel and oxygen governs reaction kinetics and thus, the information of fuel and oxygen in the mixture can be expressed by one unique variable instead of two dimensions: '*fuel*' and '*oxygen*'. In the case of the postoxidation tabulation, this assumption is no longer adapted. The equivalence ratio at the beginning of a direct injection stratified IC-engine combustion varies generally between 0.2

and 6 (due to local stratifications). However, except for the case of ideal stoichiometric oxidation, the equivalence ratio does not remain constant during the chemical reaction. Figure 7-1 shows the evolution of the equivalence ratio over the reaction progress c for a stoichiometric, a slightly poor ($\phi = 0.7$) and a slightly rich ($\phi = 1.3$) global equivalence ratios.

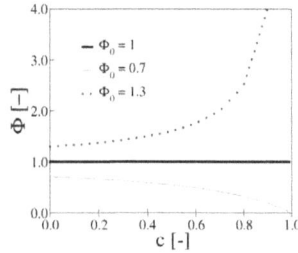

Figure 7-1   Evolution of the equivalence ratio over reaction progress c (P=1bar, T = 700 K).

It is obvious that equivalence ratios diverge towards infinity in the rich case or towards zero in the lean case. Following Figure 7-1, the equivalence ratio found at the end of reaction may vary between 0 and infinity. Such a wide range of equivalence ratios cannot be handled numerically in a FPI-table. For that reason, the thermochemical table variables of fuel and oxygen mass fractions cannot be expressed by one single parameter such as the equivalence ratio and must therefore be represented by two separate table dimensions, the oxygen and fuel mass fractions.

7.1.2. Fraction of Secondary Air and NO mass fraction

The fraction of secondary air characterizes the mass fractions of injected secondary air (SAI) in the mixture with the exhaust gases which are emitted from the IC-engine. In practice, the oxygen of the SAI favours the postoxidation of the unburned hydrocarbons. However, during lean engine combustion and even under global rich conditions, due to high stratifications, oxygen is also emitted from the engine. Thus, the oxygen issued from SAI and the oxygen emitted from the IC-engine are two independent characteristics. They cannot be expressed by one single characteristic in the FPI-table and consequently represent two separate table dimensions.

The formation of NO during engine combustion depends strongly on the temperature range at which combustion occurs (Zeldovich, 1946, Lavoie et al. 1970) and the characteristic time scale of NO formation differs from that characterizing hydrocarbon oxidation, as NO_x are produced close to reaction equilibrium, when temperatures are the highest (Flynn et al. 2000). Consequently, the NO produced during engine combustion cannot be related to the reaction progress for hydrocarbon oxidation. The NO mass fraction in the mixture must be expressed by a separate input in the FPI-table

7.1.3. Dilutant composition

It was shown in section 5.1. that the impact of the different diluting species CO, CO_2 and H_2O, compared to dilution by N_2 cannot be neglected. Combustion modelling with an a priori tabulation strategy implies that these gases have to be considered as table inputs. However, we have seen that the multiplication of table entries is not possible. A reduction strategy is therefore necessary. It is presented and explained in this section.

CO_2 and H_2O are products of a complete combustion, while CO results mainly from incomplete oxidation under rich conditions. Figure 7-2 illustrates the composition of the diluting gases at the end of combustion as a function of initial equivalence ratio. Equivalence ratios were tested between 0.2 and 6. Beyond these limits, the equivalence ratio is not a physically relevant parameter during IC-engine combustion. Results were calculated by detailed chemistry computations in a closed homogenous reactor at contant atmospheric pressure conditions for a pure fuel/air mixture ($y_{EGR} = 0$). The chosen initial temperature T_0 was 700 K. We checked the composition of reaction products at the end of reaction at various equilibrium conditions (T_{eq}, P) and found negligible variations in the residual gas composition at varying temperatures T_{eq} and pressures P. The results of these equilibrium computations on which this assumption is based are shown in appendix B.4. All results of residual gas compositions shown in the following were obtained under atmospheric pressure conditions and an initial temperature T_0 of 700 K.

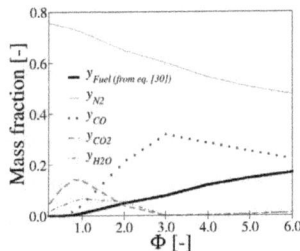

Figure 7-2   Evolution of the dilutant composition at the end of combustion as a function of the initial equivalence ratio ϕ ($P = 1$ bar, $T_0 = 700$ K).

The figure also shows the evolutions of the sum of the mass fractions of unburned fuel and partially oxidized species y_{Fuel}. These are represented by fractions of the chosen n-heptane/iso-octane/toluene fuel surrogate. The mass fraction y_{Fuel} substituting the ensemble of the unburned hydrocarbons, including CO, were obtained by determining the *lower heating value*[5] of all partially oxidized species. The lower heating value of a chemical species is calculated from its formation enthalpy h^f and its thermodynamic potential to produce CO_2 and H_2O:

$$[28] \quad q_i = \frac{\left(h_i^f - a \cdot h_{H_2O}^f - b \cdot h_{CO_2}^f\right)}{MW_i}$$

where h_i^f is the formation enthalpy of species *i*, *a* is the number of mols of H_2O and *b* is the number of mols of CO_2 produced after the complete oxidation of species *i* and MW_i is the molecular weight of the same species *i*. The mass fraction y_{Fuel} of the chosen n-heptane/iso-octane/toluene fuel surrogate corresponds to the sum of all partially oxidized hydrocarbons y_i^{UHC} and has the same lower heating value. The following equation can be written:

$$[29] \quad y_{Fuel} \cdot q_{Fuel} = \sum y_i^{UHC} \cdot q_i^{UHC} + y_{CO} \cdot q_{CO}$$

with q_{Fuel} and q_i representing the lower heating value of the fuel and species *i* respectively. This correlation allows determining the fraction of fuel which substitutes the partially oxidized hydrocarbons by keeping a constant lower heating value of the mixture. The molar fractions of the substituting fuel were thus obtained by equation [30]:

$$[30] \quad y_{Fuel} = \frac{\sum y_i^{UHC} \cdot q_i^{UHC} + y_{CO} \cdot q_{CO}}{q_{Fuel}}$$

Equation [30] includes the lower heating value of CO, which is not a hydrocarbon species. The inclusion of CO in the chosen substitution of partially oxidized species by initial fuel is necessary, as at stoichiometric and moderate rich conditions ($\phi < 3$), CO represents by far the major part of all partially oxidized species. Thus, at equivalence ratios of $\phi < 3$, the substituting fuel mass fraction y_{Fuel} is close to 0, when ignoring CO in the chosen definition [30]. The lower heating value is mainly determined by the mass fraction of CO under such stoichiometric conditions. During lean and stoichiometric combustion, a complete fuel conversion into CO_2 and H_2O is almost achieved. Under such conditions, the CO_2 and H_2O production from initial fuel is the largest and only a small amount of CO is produced. When combustion occurs under rich conditions, the production of CO increases drastically, disfavouring the formation of CO_2. The H_2O production decreases also under rich conditions due to the lack of oxygen which favours the production of partially oxidized hydrocarbons

---

[5] *The lower heating value (also known as net calorific value, net CV, or LHV) of a fuel is defined as the amount of heat released by burning a specified quantity (initially at 25 °C or another reference state) and returning the temperature of the combustion products to 150°C (Guibet, 1997).*

instead of H_2O. The results presented in Figure 7-2 reveal that the distribution of N_2, CO, CO_2 and H_2O in the diluting gases is not random and that only certain combinations of these species may physically occur. That distribution strongly depends on the equivalence ratio at which combustion occurs. The same conclusion is valid for the fractions of partially oxidized hydrocarbons in the mixture. They increase almost linearly with increasing equivalence ratio. Consequently, the fractions of the diluting compounds N_2, CO, CO_2 and H_2O may be related to the fractions of partially oxidized species. Figure 7-3 displays the mass fractions of N_2, CO, CO_2 and H_2O as a function of the partially oxidized hydrocarbons at the end of combustion. Like in Figure 7-2, the partially oxidized hydrocarbons are represented by PRF/toluene surrogate fractions having the same lower heating value.

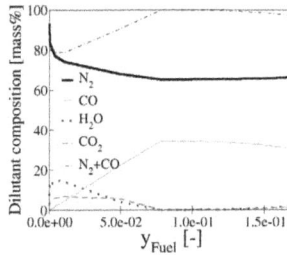

Figure 7-3 Evolution of the dilutant composition at the end of combustion as a function of the fractions of partially oxidized hydrocarbon. (P = 1 bar, T_0 = 700 K).

Figure 7-3 illustrates the correlation between N_2, CO, CO_2 and H_2O evolutions and the fraction of partially oxidized hydrocarbons (y_{Fuel}). At low hydrocarbon mass fractions in the burned gases (corresponding to lean and stoichiometric combustion conditions), the CO_2 and H_2O mass fractions vary between 0 and 15 mass% and 10 mass% respectively, while CO mass fractions do not exceed 10 mass%. With increasing hydrocarbon mass fractions (increasing equivalence ratio), the CO mass fractions increase up to ~35 mass% and CO_2 and H_2O fractions fall down to far less than 1 mass% showing a minimum at 7 mass% of unburned hydrocarbons (corresponding to an equivalence ratio ϕ of ~3). For higher fractions of unburned hydrocarbons ($\phi > 3$), the CO_2 and H_2O increase again to up to 1 mass%. The observed increase of CO_2 and H_2O mass fractions at very high equivalence ratios should be interpreted with care as the reaction mechanism has not been validated for these stoichiometries. However, the fractions of CO remain greater than 30 mass% at any fraction of substitution fuel above 7 mass%. N_2, CO_2 and H_2O are thus related to the amount of partially oxidized hydrocarbons and correlations can therefore be written for the FPI-table generation according to the evolutions plotted in Figure 7-3:

$$[31] \quad y_{N_2} = f(y_{Fuel})$$

$$[32] \quad y_{CO_2} = f(y_{Fuel})$$

$$[33] \quad y_{H_2O} = f(y_{Fuel})$$

No corresponding correlation was defined for the concentrations of CO, even if this was feasible. The proposed approach of linking H_2O fractions to the lower heating value of unburned hydrocarbons should be good over the widest range of conditions, including slightly rich and stoichiometric conditions. At slightly lean to moderate rich conditions the major proportion of the burned gases lower heating value results from CO. Thus, CO is considered not as dilutant, but as partially oxidized compound. Consequently CO is substituted by fuel with an equivalent LHV and not considered in the dilutant composition.

With these assumptions, the chosen tabulated inputs for generating the FPI-table are the following six dimensions:

1. Fraction of Secondary Air: SAI_0
2. Pressure: P_0
3. Temperature: T_0
4. NO mass fraction: y_{NO_0}
5. Fuel mass fraction: y_{CxHy_0}
6. Oxygen mass fraction: y_{o2_0}

7.2   Definition of the progress variable, transported species and table discretization

This section describes the chosen definition of the reaction progress variable and the reconstructed and transported species. The last part of this section explains the table discretisation chosen for its different dimensions (SAI_0, P_0, T_0, y_{NO_0}, y_{CxHy_0}, y_{o2_0}).

7.2.1. Definition of the progress variable

The importance and principles for building the progress variable c for FPI-tabulations were explained in section 6.1.4. We have chosen the commonly definition of c from CO and CO_2 as described in equations [16] and [17] (see appendix B.1). At reaction start ($c = 0$), the concentration of the combustion product CO is set to 0, as CO is included in the reconstruction of unburned hydrocarbons by y_{Fuel}^{init}.

As the progress variable is the main link between the tables and the CFD code, it is important to detail all the transition from one to the other. Figure 7-4 shows a principle sketch of the coupling interactions between the developed FPI-table, and the CFD-code *IFP-C3D*. The i species mass fractions $y_{j,i}(t)$ of the tabulated species ($y_{j,CxHy}$, $y_{j,O2}$, $y_{j,CO}$, $y_{j,CO2}$, $y_{j,H2O}$, $y_{j,H}$) plus the three species mass fractions $y_{j,i}^{rec}(t)$ of the reconstructed species ($y_{j,C7H8}$, $y_{j,H2}$, $y_{j,OOH}$) and the mass fraction of NO $y_{j,NO}$ are transported by the CFD-code. For each computational cell j, the progress variable $c_j(t)$ is computed at time t by equations [16] and [17] based on the transported mass fractions y of species CO and CO$_2$. The trajectory corresponding to the thermodynamic condition in cell j is located in the FPI-table via the table dimensions SAI_0, P_0, T_0, y_{NO}, y_{CxHy0} and y_{O20}. Depending on $c_j(t)$, the mass fractions $y_{j,i}(t)$ of each tabulated species are advanced according to equations [19]-[23] and the tabulated mass fractions $y_{j,i}^{tab}(t+dt)$ are computed. Mass and element balances are closed by re-computation of the reconstructed species mass fractions $y_{j,i}^{rec}(t+dt)$ from equations [11]-[13]. The updated mass fractions $y_{j,i}(t+dt)$ are given to the CFD code, the thermodynamic conditions ($P_j(t+dt)$, $T_j(t+dt)$) are re-calculated as a function of the heat release during dt obtained via equation [14].

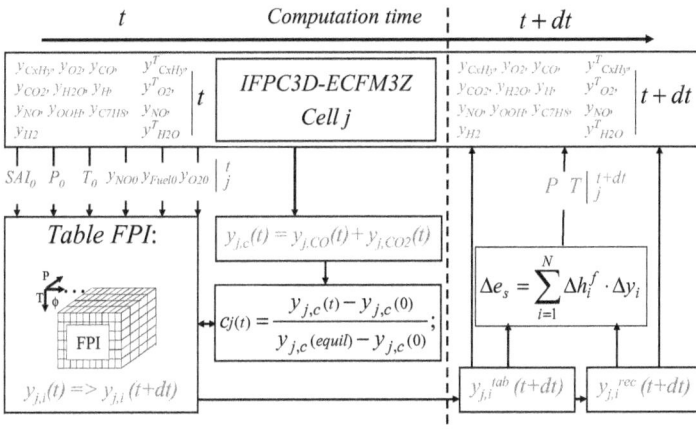

Figure 7-4   Scheme of the coupling between the new developed FPI-table valid at postoxidation conditions and the CFD-code IFP3CD.

7.2.2. Transported and reconstructed species

The following species mass fractions are tabulated in the FPI-table derived for postoxidation applications:

- y_{CxHy},   - y_{O2} - y_{CO}

- y_{CO2} - y_{H2O},   - y_H

These species represent the initial fuel, oxygen and the main combustion products. Atomic H has also been tabulated in order to correctly balance the internal energy. For closing atomic C-, H- and O- balances we have reconstructed the following 'dummy'-species:

- y_{OOH} - y_{H2} - y_{C7H8}

For the purpose of the postoxidation model, the O_2 mass fractions have to be tabulated. When O-balances are closed by O_2, as it is proposed by Pera et al. (2009), important errors in O_2 concentrations may arise under rich conditions. For that reason, O-balances must not be closed by O_2 but rather by another species like OOH. Their mass fractions are deduced from the following atomic balance equations:

$$[34] \quad y_{HO_2} = \frac{M_{HO_2}}{2}\left[2 \cdot \frac{y^T_{O_2}}{M_{O_2}} - \left(\frac{y_{CO}}{M_{CO}} + 2 \cdot \frac{y_{CO_2}}{M_{CO_2}} + \frac{y_{H_2O}}{M_{H_2O}} + 2 \cdot \frac{y_{O_2}}{M_{O_2}} \right) \right]$$

$$[35] \quad y_{C_xH_y} = \frac{M_{C_xH_y}}{7}\left[x \cdot \frac{y^T_{C_xH_y}}{M_{C_xH_y}} - \left(x \cdot \frac{y_{C_xH_y}}{M_{C_xH_y}} + \frac{y_{CO}}{M_{CO}} + \frac{y_{CO_2}}{M_{CO_2}} \right) \right]$$

$$[36] \quad y_{H_2} = \frac{M_{H_2}}{2}\left[y \cdot \frac{y^T_{C_xH_y}}{M_{C_xH_y}} + 2 \cdot \frac{y^T_{H_2O}}{M_{H_2O}} - \left(y \cdot \frac{y_{C_xH_y}}{M_{C_xH_y}} + 8 \cdot \frac{y_{C_xH_x}}{M_{C_xH_x}} + 2 \cdot \frac{y_{H_2O}}{M_{H_2O}} + \frac{y_H}{M_H} \right) \right]$$

where y_i is the mass fraction of species i, M_i its molar weight, and y_i^T is the concentration of the tracer species i known from the ECFM-3Z model. A separate closure of N atoms is not necessary as N_2 and NO are considered as residual species. The presence of NO strongly impacts hydrocarbon oxidation kinetics (see chapters 2 and 5). Nevertheless, kinetic analyses revealed that under postoxidation conditions, the NO concentrations remain almost constant and that the impact of NO is similar to that of a catalyst. However, the concentrations of NO and N_2 have to be known and must hence be transported in the IFP-C3D code. Thus, the following 11 species mass fractions are transported:

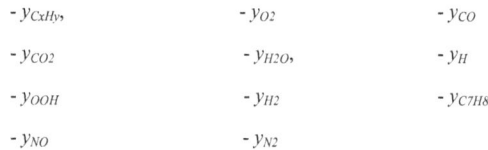

- y_{CxHy},   - y_{O2} - y_{CO}

- y_{CO2} - y_{H2O},   - y_H

- y_{OOH} - y_{H2} - y_{C7H8}

- y_{NO} - y_{N2}

Our choice of tabulated and reconstructed species differs from that made by Pera et al. (2009) who tabulated five species (C_7H_{16}, CO, CO_2, H_2O and H) and reconstructed the atomic C, H, and O-balances with the help of 'dummy'-species C_7H_{14}, O_2, and H_2. The choice made by Pera et al. (2009) for reconstructing C- and O- balances by C_7H_{14}, H_2 and O_2 leads to several difficulties:

- The H/C molar ratio of C_7H_{14} is 2 which is greater than the average H/C ratio of partial oxidized components in rich conditions, when the equivalence ratio ϕ exceeds 3. In that case, the calculated H_2 concentrations for closing the H- balance may become negative.
- Under rich conditions, the atomic oxygen fraction in the partially oxidized components becomes important. This results in a dramatic overestimation of O_2 concentration in rich conditions leading to overlean mixtures.

Our choice of tabulating O_2 and closing the C-balance by a C_7H_8 species avoids this problem.

7.2.3. Table discretization

Several constraints govern the look-up table dimensions. Table input discretization must be accurate enough so that errors resulting from linear interpolations between table points are minimized. On the other hand, the total number of table points should not exceed a certain value because of computational memory limitations. In section 7.1, we have determined six independent table dimensions SAI_0, P_0, T_0, NO_0, $Fuel_0$ and O_{2_0}. The discretization and the table range of each table input are described in the following section.

7.2.3.1. Fraction of injected secondary air (SAI_0)

The fraction of injected secondary air is defined as the fraction of O_2 and N_2 (in mass percent) issuing from SAI, over the total burned gas mass. The fraction of SAI α is defined following equation [24]:

$$[24] \qquad \alpha = y_{N_2}^{SAI} \cdot \left(1 + \frac{0.23}{0.77}\right)$$

where y_{N2}^{SAI} denotes for the mass fraction of N_2 issued from the injected air. The ratio 0.23/0.77 is the mass ratio of N_2 and O_2 in air. Figure 7-5 illustrates the correlation between, burned gas composition and the definition of the SAI fraction α.

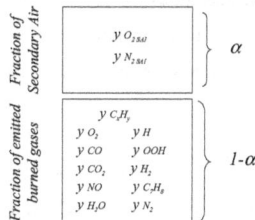

Figure 7-5   Definition of the fraction α of injected secondary air (SAI) and the burned gas fraction (1-α) in a computational cell.

The values of α may thus vary between 0 and 1. Close to the combustion chamber exhaust valve, only burned gases are present ($\alpha = 0$) while in the local neighbourhood of the SAI system input, the mixture will consist of only secondary air ($\alpha = 1$). Burned gases may also react in absence of secondary air ($\alpha = 0$) if oxygen resulting from a lean combustion is present in the burned gases and mixes with UHC leaving crevices and the lubricant film. The sensitivity analysis on mixture reactivity shown in appendix B.3 revealed that PRF/toluene mixtures are the most reactive in rich conditions. For that reason, low α values were thoroughly discretized. Eight values of α have been chosen. They are listed in Table 26.

7.2.3.2. Pressure (P_0) and Temperature (T_0)

While pressure in the exhaust line of an IC-engine is mostly around the atmospheric pressure (application without turbocharger), the pressure evolution shown in Figure 1-3 reveals that the pressure in the different computational areas where postoxidation takes place may differ significantly from the atmospheric pressure. Pressure was thus discretized by two values, 0.5 and 2 atm. Figure 1-4 shows that exhaust gas temperatures may vary between 600 and 1100 K. We tested reaction kinetics of highly diluted mixtures at temperatures below 600 K and could not detect noticeable reaction kinetics. Therefore, we have fixed the minimum tabulated temperature at 600 K. The maximum temperature was chosen to 1200K and we descretized 20 temperature values in intevalls of 25 K (T < 1000 K) and 50 K (T > 1000 K).

7.2.3.3. Mass fraction of NO (NO_0)

It was shown that hydrocarbon oxidation is very sensitive to the presence of NO (appendix B.3). A complex impact of NO depending on temperature, pressure and NO concentration has been reported in chapters 2, and 4. At atmospheric pressure, NO has an accelerating impact at concentrations as low as 50 ppm (Moréac, 2003). At higher concentrations (NO > 200 ppm), NO showed an inhibiting impact at low temperatures (T < 750 K) while at higher temperatures, an accelerating impact on hydrocarbon oxidation can be attributed to NO (Moréac, 2003; Dubreuil et al. 2005). In order to cover all potential impact of NO on hydrocarbon combustion, we have tabulated four NO mass fractions of 0, 5×10^{-5}, 5×10^{-4}, 1.5×10^{-3}.

7.2.3.4. Mass fraction of fuel (y_{CxHy_0})

It was shown that unburned hydrocarbons and CO can be replaced by a PRF/toluene fuel surrogate with a heat of combustion identical to that of the unburned hydrocarbons and CO mixture. By doing so, the mass fractions of the substituting fuel may rise up to 17 masse% at the largest equivalence ratio ($\phi = 6$, see the mass of to the sum of UHC and CO mass fractions plotted in Figure 7-2). These mass fractions were computed by substituting the fuel surrogate as a function of the equivalence ratio, obtained by calculation of the fuel mole fractions following equation [30] and by converting them into mass fractions. In average, the high hydrocarbon levels in the exhaust gases computed for high equivalence ratios are not common in IC-engine applications since the maximum global equivalence ratio in a spark ignition engine is about 1.2 for heavy load. However, such extreme conditions may appear locally from resulting mixture stratification in the combustion chamber or the exhaust manifold.

According to section 7.1.3, the distribution of the diluting compounds N_2, CO_2 and H_2O in the exhaust gases is a function of the fuel. The choice of the tabulated fuel mass fractions has to take this into account. Figure 7-6 displays the dilutant composition as a function of the substituting fuel surrogate. It can be seen that the most important variations are located in a region where the fuel mass fraction is around 0.1. Discretized fuel concentrations are thus concentrated around this value. They are represented by the vertical dotted lines in Figure 7-6.

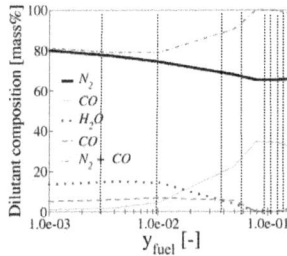

Figure 7-6   Dilutant composition of the burned gases in mass % as a function of the substituting fuel surrogate calculated for fresh gas oxidations at equivalence ratios ϕ varying from 0.2 to 6. Discretized fuel mass fractions are indicated by the vertical dotted lines ($P = 1$ bar, $T_0 = 700$ K).

A minimum fuel mass fraction of 10^{-3} was chosen. This value corresponds to the heat of reaction of an exhaust gas produced from fresh gas oxidation at an equivalence ratio between 0.6 and 0.7. For lower equivalence ratios, the heat of reaction of the produced exhaust gases is negligible and hence no postoxidation activity is expected under such conditions. Ten fuel mass fractions were tabulated up to a maximum fuel mass fraction of 0.17. The tabulated fuel mass fractions are listed in Table 26.

7.2.3.5. Mass fraction of Oxygen (O_{2_0})

The excess of oxygen in the engine exhaust gases results from a lean combustion or from stratifications inside the combustion chamber. The excess of oxygen is related to the global equivalence ratio at which fresh gases have been burned. Figure 7-7 shows the mass fractions of excess oxygen as a function of the global equivalence ratio at which the fresh gases were oxidized.

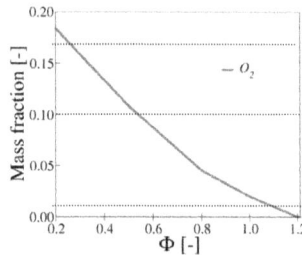

Figure 7-7   Mass fraction of excess oxygen computed by the developed detailed reaction mechanism for the oxidation of a pure fuel/air mixture (EGR=0) for equivalent ratios ϕ varying from very lean ($\phi=0.2$) to slightly rich conditions ($\phi=1.2$). Horizontal lines mark the discretized mass fractions of excess oxygen ($P = 1$ bar, $T_0 = 700$ K).

Under very lean conditions ($\phi = 0.2$), up to 17 mass% of the exhaust gases consists of oxygen. With increasing equivalence ratio, the fractions of excess oxygen decrease towards zero for an equivalence ratio around 1.2. Figure 7-7 clearly shows the correlation between the excess oxygen in the burned gases and the equivalence ratio at which fresh gases were burned. Hence, a correlation between excess of oxygen and the substituting fuel could be established, following a strategy similar to the one described for the dilutant composition in the previous section. However, in that case, the fact that fuel might evaporate from crevices into zones of excess oxygen would be neglected. Also, in a stratified engine, regions containing either excess oxygen or unburned fuel may exist in the exhaust gases. Thus, we have chosen not to correlate the excess of oxygen with the unburned hydrocarbons contained in the exhaust gases, but to tabulate its values in an independent table dimension. Four excess oxygen mass fractions were tabulated: 0, 0.01, 0.1 and 0.17 (horizontal lines on Figure 7-7).

7.2.4. Overview of the table dimensions and table size

Table 25 shows the table inputs and the values chosen for the table. The dilutant composition is correlated to the tabulated fuel mass fractions according to Figure 7-2 and Figure 7-3.

Table 25   Dimensions and discretization values of generated FPI-table for postoxidation conditions.

Table Dimensions						Dilutant composition (in function of Fuel)		
α	P	T	NO	O_2	Fuel	H_2O	CO_2	N_2
[-]	[atm]	[K]	[kg/kg]	[kg/kg]	[kg/kg]	[mass%]	[mass%]	[mass%]
0	0.5	600, 625	0	0	0.001	5.0	13.0	82.0
0.001	2	650, 675	5.0×10^{-5}	0.01	0.003	6.0	14.0	80.0
0.005		700, 725	5.0×10^{-4}	0.1	0.01	7.0	15.0	78.0
0.01		750, 775		0.17	0.05	7.0	4.0	89.0
0.1		800, 825			0.07	1.0	1.0	98.0
0.2		850, 875			0.1	0.5	0.1	99.4
0.5		900, 925			0.12	1.0	0.1	98.9
0.8		950, 975			0.13	1.0	0.1	98.9
		1000, 1050			0.15	1.0	0.1	98.9
		1100, 1200			0.17	2.0	0.1	97.9

For each table set of conditions, the progress variable c has been discretized by 36 values between 0 and 1. Thus, the generated table consists of 38 400 computed thermochemical conditions and 1 382 400 mixture compositions have to be stored.

The FPI-table is read as a function of the thermochemical conditions in the computational area discretizing the IC-engine combustion chamber and its exhaust line. That state is defined by the different table dimensions. Figure 7-8 illustrates the table dimensions defining the thermochemical state for which reaction manifolds are read. The procedure of table generation is described in appendix B.5.

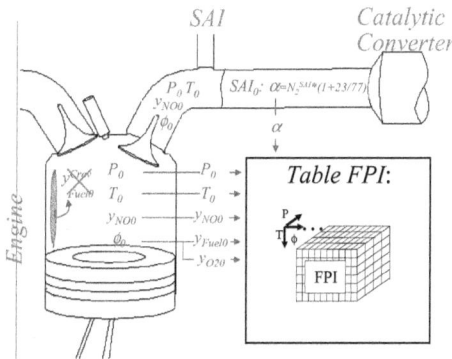

Figure 7-8   Table dimensions characterizing the thermochemical conditions in the computational area inside the combustion chamber and the IC-engine exhaust line.

The FPI-model developed in this work defines the thermochemical state by the properties of secondary air injected into the exhaust line α, the pressure and temperature P_0, T_0, and the mass fractions y_{NO}, y_{Fuel}, y_{O2}. It was shown that the mass fractions of fuel $y_{Fuel}^{Crevices}$ issued from crevices should have a determining impact on mixture reactivity. However, such fuel mass fraction $y_{Fuel}^{Crevices}$ is not considered in the present model. The implementation of that additional table dimension will be part of a model extension foreseen in future work.

7.3   Validation against detailed chemistry

The postoxidation FPI model strategy described in the previous sections was built from complex chemistry simulations and implemented in the IFP-C3D software. Linear interpolation algorithms were also written and implemented in the code in order to retrieve data between tabulated points. This section describes a priori model validations where the IFP-C3D results were directly compared with complex chemistry simulations performed with the Chemkin 4 software package. A very simple geometric configuration was built in IFP-C3D in order to cope with the fact that Chemkin only deals with simple geometries. A mesh of 32 rectangular cells was then generated representing a constant volume closed homogenous reactor. For validation purposes, a reduced table was generated containing only two values in each dimension except α (the fraction of secondary air) which was set to 0 since no air injection may occur in the closed system. Table 26 displays the table dimensions and their discretization points.

Table 26   Table dimensions and discretization values of the generated test table.

Table Dimensions						Dilutant composition (Function of Fuel)	
α	P	T	y_{NO}	y_{Fuel}	y_{O2}	y_{H2O}	y_{N2}
[-]	[bar]	[K]	[-]	[-]	[-]	[mass %]	[mass %]
0	1.5	1050	0	0.01	0.1	7.0	93.0
0	2	1100	5.0×10^{-5}	0.05	0.15	7.0	93.0

We have first defined a reference case at a pressure of 2 bar, temperature of 1100 K, NO mass fraction of $5.\times 10^{-5}$, fuel mass fraction of 0.01 and oxygen mass fraction of 0.1. The initial conditions of the reference case correspond exactly to the conditions of one tabulated point of the FPI table. Thus, for this case, no interpolations were performed when reading the table and the mass fractions read in it match exactly those computed by CHEMKIN during table generation. For each table dimension P, T, y_{NO}, y_{Fuel} and y_{O2}, two additional computations were performed by varying the initial conditions along the tested dimension. One corresponds to the other tabulated point, and the other to the middle value between the two tabulated values. This procedure allows validation of the following items:

- The manifolds corresponding to the thermochemical conditions are located and read correctly.
- Conditions located between two tabulated points are interpolated correctly.

Computations were performed by the *IFPC3D* CFD-code and compared to the results obtained by corresponding CHEMKIN computations. Initial conditions of all computations performed are listed in Table 27.

Table 27   *Computations performed in IFP-C3D for the validation of the tabulated chemistry against closed homogenous reactor simulations performed with Chemkin.*

Validation	α	P	T	y_{NO}	y_{Fuel}	y_{O2}
	[-]	[bar]	[K]	[-]	[-]	[-]
Ref	0	2	1100	0	0.01	0.1
P_1	0	1.75	1100	0	0.01	0.1
P_2	0	1.5	1100	0	0.01	0.1
T_1	0	2	1075	0	0.01	0.1
T_2	0	2	1050	0	0.01	0.1
NO_1	0	2	1100	2.5×10^{-5}	0.01	0.1
NO_2	0	2	1100	5.0×10^{-5}	0.01	0.1
Fuel_1	0	2	1100	0	0.03	0.1
Fuel_2	0	2	1100	0	0.05	0.1
Oxy_1	0	2	1100	0	0.01	0.125
Oxy_2	0	2	1100	0	0.01	0.15

Figure 7-9 compares the results obtained with IFP-C3D with those obtained by CHEMKIN for conditions listed in Table 27. For points corresponding exactly to tabulated values, the IFP-C3D computed results are superposed with those simulated with CHEMKIN. IFP-C3D computed temperature evolutions may deviate slightly from the temperatures calculated by CHEMKIN due to species reconstruction by mass balances in IFP-C3D. However, CFD computed ignition delays agree exactly with those calculated with CHEMKIN. This shows that the FPI table is correctly built and implemented in the CFD code.

When input values are interpolated (points of type 1 in Table 27), IFP-C3D computed ignition delays are also very close to those determined with CHEMKIN when the pressure is changed (Figure 7-9a). This is due to small interpolation errors. They can however be minimized by refining the discretization of the corresponding table dimension. In our case, the order of magnitude of interpolation errors is perfectly acceptable.

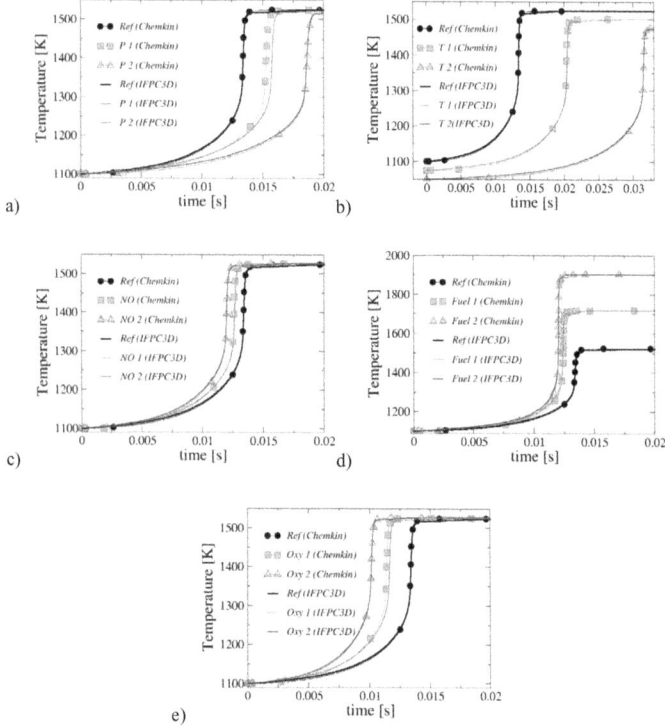

Figure 7-9   *Comparison of IFP-C3D computation results in a closed homogenous system (lines without symbols) with CHEMKIN 4 simulations of the same reactor geometry (lines with attached symbols): a) Pressure P variation, b) Temperature T variation, c) NO mass fraction y_{NO} variation, d) Initial fuel mass fraction y_{Fuel} variation, e) Oxygen mass fraction y_{Oxygen} variation.*

7.4   Reinitialisation of the composition field

The presented FPI-tabulation methodology was developed for CFD-postoxidation computations of burned gases produced by in-cylinder combustion. However, since the postoxydation combustion model is not adapted for high temperature, variable pressure computations, a transition between the two models is necessary. This section explains the need of reinitialisation of the composition field between the end of IC-engine computations and the beginning of postoxidation simulations. Then, the reinitialisation equations are shown.

7.4.1. Necessity for flow field composition reinitialisation

One of the key aspects of the FPI model for postoxidation applications developed in this work is the reinitialisation of the composition space at the moment when IC-engine main combustion ends and postoxidation starts. This operation is necessary because during the transition from main combustion (IC-engine) to postoxidation regimes (computations in the exhaust line), simulations switch between different combustion models. IC-engine combustion computations are performed by ECFM3Z (Figure 6-1) while postoxidation computations are based on the developed FPI-approach. Since the latter model is based on kinetic tabulations with specific inputs, the flow field must accommodate these entries.

At the end of IC-engine combustion, reaction progress is equal or close to unity and important fractions of the burned gases are close to their thermodynamic equilibrium. By opening the exhaust valves and then eventually injecting secondary air into the burned gases, the thermodynamic equilibrium of such a mixture strongly changes and kinetic reactions (postoxidation) may occur. Thus, the mixture reaction states have to be reset to zero at the beginning of the postoxidation process in order to account for the shift in thermodynamic equilibrium. Precise information on initial thermodynamic conditions as well as on mixture composition is crucial for correctly reading the FPI-table. However, the composition of the computed exhaust gases after IC-engine oxidation is defined in such a way that it cannot be directly used by the FPI postoxidation model (see section 6.2.2). Consequently the exhaust gas composition at the beginning of postoxidation processes has to be reconstructed, together with the reset of the progress variable. Such reconstruction must be performed from the information available after IC-engine computations.

When tabulating chemistry, a choice of a surrogate fuel able to represent primary fuel as well as UHC was made. Complex kinetic simulations will therefore be based on that fuel. However, depending on the main combustion process and on the corresponding model, UHC composition might not be coherent with the postoxidation surrogate fuel neither in terms of C and H atoms balance nor available enthalpy. When dealing with tabulation models based on species, a perfect element or enthalpy match between the table and the CFD computation has to be ensured. This is not granted if the initial postoxidation CFD computation fuel is different from the tabulated one. For that reason a numerical switch between both combustion models is necessary. A transition moment for switching between the combustion models also has to be defined. This could be the moment of exhaust valve opening, when the exhaust line becomes connected to the combustion chamber. It should be noted that the study of the transition moment between the two modelling approaches (main combustion and postoxidation) was not the object of this work and has not been treated here. This should be the object of future work.

7.4.2. Reconstruction of the composition space by energy balances

When closing molar and element balances, mass is conserved. Nevertheless, reconstruction of the composition space by element balances implies some major disadvantages, as such a method is not conservative in energy. This might lead to important differences in the final temperature of the combustion products. The problem of reconstructing the composition space by element balances is discussed in detail in appendix B.6. We propose to define the initial exhaust gas composition by conserving energy rather than mass. The main input for postoxidation is the lower heating value left over in the gases. Due to the high degree of dilution, a reactive system under postoxidation conditions is rather insensitive to small inaccuracies in mass conservation, but depends strongly on the lower heating value of the gas mixture. The lower heating value is therefore input of the first order to be kept when the exhaust gases composition space is reconstructed.

It was mentioned that the reconstructed species accumulated during the IFP-C3D computations during IC-engine combustion represent all partially oxidized compounds which were not transformed into CO, CO_2 or H_2O. The composition of the partially oxidized hydrocarbons contained in the burned gases close to reaction equilibrium was determined with the help of the developed reaction mechanism as a function of equivalence ratio and is displayed in Figure 7-10. Species mass fractions are shown for CO and the major partially oxidized compounds: benzene, methane (CH_4), ethylene (C_2H_4) and acetylene (C_2H_2).

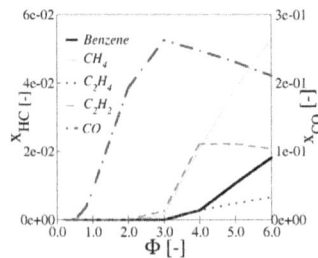

Figure 7-10   Mol fractions of CO and partially oxidized hydrocarbons as at the end of combustion as function of equivalence ratio φ at which the fresh gases were burned. (P = 1 bar, T_0 = 700 K)

The species shown in Figure 7-10 represent the major compounds contributing to the lower heating value of the burned gases after fresh gas oxidation. Lower heating values related to other species are negligible due to their low concentrations close to reaction equilibrium. Table 28 presents the formation enthalpies, oxidation coefficients a and b as well as the calculated lower heating value q following equation [28]:

$$[28] \quad q_i = \frac{\left(h_i^f - a \cdot h_{H_2O}^f - b \cdot h_{CO_2}^f\right)}{MW_i}$$

The lower heating value of the fuel surrogate was calculated by equation [37]:

$$[37] \quad q_{Fuel} = 0.137 \cdot q_{C_7H_{16}} + 0.429 \cdot q_{C_8H_{18}} + 0.434 \cdot q_{C_7H_8}$$

All listed formation enthalpies correspond to those published in the NIST-database.

Table 28   Formation enthalpies and lower heating values of the initial fuel components and the major species present in the partially oxidized compounds.

Species	h_i^f [kJ/mol]	a [-]	b [-]	q_i [kJ/mol]	q_i^{mass} [kJ/g]
Fuel	-100.2	6.693	7.429	4329.9	42.2
n-heptane	-187.8	8	7	4652.8	46.4
iso-octane	-224.1	9	8	5251.8	46.0
toluene	50	4	7	3316.6	36.0
CO	-110.5	0	1	131.3	4.7
benzene	82.8	3	6	2714.1	34.7
CH_4	-74.8	2	1	954.0	59.5
C_2H_4	52.47	2	2	1323.1	47.2
C_2H_2	226.7	1	2	1103.8	42.4
H_2O	-241.8	0	0	0	0
CO_2	-393.5-	0	0	0	0

Values in Table 28 allow the computation of the total lower heating value contained in the burned gases q_{BG} for any state of composition:

$$[38] \quad q_{BG} = x_{Fuel} \cdot q_{Fuel} + x_{CO} \cdot q_{CO} + x_{rec\,spec} \cdot q_{rec\,spec}$$

with x_{Fuel} as the mol fraction of unburned fuel, x_{CO} the mol fraction of CO and $x_{rec\,spec}$ the mol fraction of the reconstructed species. q indicates the lower heating value referenced in Table 28 for the corresponding species. $q_{rec\,spec}$ corresponds to values of the reconstructed species and are obtained by equation [39]:

$$[39] \quad q_{rec\,spec} = x_{CH_4} \cdot q_{CH_4} + x_{benzene} \cdot q_{benzene} + x_{C_2H_2} \cdot q_{C_2H_2} + x_{C_2H_4} \cdot q_{C_2H_4}$$

As outlined in Figure 7-11, the lower heating value of the reconstructed species $q_{rec\,spec}$ in the burned gases is a function of the fresh gas equivalence ratio. The heating values plotted as a function of the equivalence ratio ϕ .are those shown already in Figure 7-10.

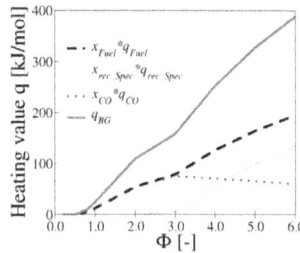

Figure 7-11   Heating values q of the reconstructed species (including CO), CO and the unburned hydrocarbons (UHC) contained in the burned gases at the end of combustion as function of equivalence ratio φ at which the fresh gases were burned.(P = 1 bar, T₀ = 700 K)

Correlating the gas composition at the end of reaction to the stoichiometry at which fresh gases reacted, allows a physically correct estimation of the energy contained in the burned gases and computed by IFP-C3D during IC-engine combustion. Along the process, the fractions of unburned fuel and CO are read in a FPI-table and thus well defined. No information on the composition of partially oxidized hydrocarbons is given beyond the fractions of the lumped reconstructed species. However, in IFP-C3D, the equivalence ratio ϕ at which fresh gases were burned is known as well. This information is stored in tracers tracking virtually unreactive initial oxygen and fuel which are transported in the same manner as the reactive physical species. The knowledge of the initial equivalence ratio allows the reconstruction of the realistic composition of the reconstructed species as it was shown in Figure 7-10. Once the global composition of the reconstructed species has been estimated, the global lower heating value of the gas composition can be calculated by equations [38] and [39]. After reinitialisation of the species composition space, the global lower heating value in the burned gases is totally represented by the fuel surrogate. Hence, the fuel mass fraction is calculated by:

$$[40] \quad y_{Fuel}^{init} = \frac{q_{BG}/W_{BG}}{q_{Fuel}/W_{Fuel}}$$

In section 7.1, it was explained that the postoxidation model correlates the dilutant composition with the fuel mass fraction:

$$[41] \quad y_{CO_2}^{init} = f(y_{Fuel}^{init})$$

$$[42] \quad y_{H_2O}^{init} = f(y_{Fuel}^{init})$$

The CO lower heating value was considered in the calculations of the mixture lower heating value and hence, the initialized mass fractions of CO are 0.

$$[43] \quad y_{CO}^{init} = 0$$

In our model, the initial O_2 mass fraction is equal to the one computed at the end of the IC-engine calculation:

$$[44] \qquad y_{O_2}^{init} = y_{O_2}^{IFPC3D}$$

$$[45] \qquad y_{NO}^{init} = y_{NO}^{IFPC3D}$$

Finally the mass fractions of the initialized composition are closed by N_2.

$$[46] \qquad y_{N_2}^{init} = 1 - y_{Fuel}^{init} - y_{O_2}^{init} - y_{H_2O}^{init} - y_{CO_2}^{init} - y_{NO}^{init}$$

When the composition space is re-initialized, pressure and temperature are kept constant. The fraction of injected secondary air α is equal to zero in the combustion chamber when postoxidation starts. If an engine plus exhaust system were to be computed, the re-initialisation procedure could be started at exhaust valve opening and the described reconstruction of the composition space applied to the complete computational domain, (exhaust line connected to the combustion chamber).

7.5   Model restrictions and potential extensions

The developed FPI-model was designed for the reproduction and understanding of postoxidation processes in the burned gases of IC-engines. It was explained that the major challenge for modelling postoxidation conditions is the reconstruction of the burned gas composition at which postoxidation starts.

7.5.1. Model restrictions

Our modelling approach implies the following major assumptions:

• *The lower heating value is estimated for the burned gases computed by IFP-C3D for IC-engine combustion.*

The IFP-C3D computations for IC-engine combustion using the FPI model are conservative in mass. However, differences in the total free enthalpy compared to the tabulated chemistry data are accumulated because of species reconstruction in the 3D code. When the burned gas composition space is re-initialized before the start of postoxidation calculations, the lower heating value contained in the burned gases is re-estimated. However, this operation is not performed by a strict closure of an enthalpy balance. Thus, the postoxidation computations are neither mass-conservative, nor energy conservative compared to the initial IC-engine computation conditions.

- *No distinction is made between evaporated fuel from walls and unburned hydrocarbons which were left over from a rich combustion.*

  This assumption is a model limitation. Figure 6-8 showed that during the oxidation of a multi-component fuel, the ratio between alkanes and aromatics is not constant and that close to reaction equilibrium, the remaining fuel consists primarily of aromatic compounds. This fact is not taken into account in our model where the fuel composition is not one of the inputs. The impact on reactivity of replacing the real fuel by a fixed mixture was tested and the results were shown in Figure 6-9. We justified the substitution of unburned hydrocarbons by the fuel surrogate with the argument that most of the unburned hydrocarbons emitted from a spark ignition engine are released from crevices during the expansion stroke.

- *The dilutant composition which is related to the fuel present in the burned gases is that obtained for reaction equilibrium.*

  At the end of IC-engine combustion, before the exhaust valve opens, not all the reacting gases are at their reaction equilibrium. Due to stratification and especially to piston expansion, reactions may be 'frozen' and large parts of the gases are not completely oxidized so that reaction progress c is less than one. In that case, the dilutant composition does not correspond to the one correlated with the substituting fuel mass fraction. This assumption may become important in lean conditions, when no ignition has occurred. In that case, important amounts of H_2O and CO_2 come from the dilutant since no oxidation has occurred. However, under rich conditions, this error is negligible as H_2O and CO_2 concentrations are very small even at reaction equilibrium. CO, which is produced in large amounts under rich conditions, is not considered in the dilutant. Hence, the dilutant at reaction start and reaction equilibrium mainly consists of N_2. Consequently, the dilution compositions are correctly reproduced under any engine condition, except for non-reacting mixtures at stoichiometric and poor stoichiometry.

7.5.2. Potential model extensions

The major improvement that is foreseen for the proposed model is the introduction of information about fuel desorbed form combustion chamber crevices. This information is important for a better reproduction of postoxidation kinetics in IC-engines and will also be useful if dealing with fuel injected during the exhaust stroke or inside the exhaust duct. Similar to the distinction between primary combustion air and secondary air, it would be necessary to distinguish between unburned hydrocarbons formed during combustion and fuel entering the combustion chamber after combustion. The FPI-table for postoxidation might then be built as follows:

- The new injected fuel represents an extra table dimension.

- The partially oxidized fuel can be represented by a mixture of methane and benzene (or toluene as well), two of the major unburned hydrocarbons directly produced by combustion.

Such model takes into account both types of fuels, the kinetically reactive original fuel and the kinetically less reactive partially oxidized hydrocarbons. Methane and toluene behave kinetically similar to the partially oxidized hydrocarbons, which mainly consist of methane and aromatic compounds. Thus, all major thermochemical impacts characterizing postoxidation kinetics could be taken into account and a better model predictivity for IC-engine postoxidation computations might be expected. However, the integration of one further table dimension would increase its number to 7 and table sizes up to 10^5 tabulated conditions can be expected. We performed tests with tables of comparable size and showed that such tables can be handled by the IFP-C3D software.

Chapitre 8

CFD-Simulation

The FPI postoxidation model developed in this study was implemented and tested in the engine 3D CFD code IFP-C3D. In the previous chapter, the model details were discussed and it was tested in a very simple geometric configuration. The objectives there were to evaluate the model assumptions, the implementation and interpolation algorithms and its ability to reproduce complex kinetic simulation results. In the present chapter, postoxidation chemistry is simulated in a system aimed to represent a real engine exhaust pipe where postoxidation chemistry may occur. In order to test the developed model only, and for the sake of simplicity, the flow in an axisymmetrical cylindrical pipe open at both ends was considered. We assumed set as well mass fractions of initial CO_2 to 0 and assumed inert gases to consist of NO, N_2 and H_2O only. The input section is divided into a central duct where fresh air is injected (like secondary air in an engine) and a coaxial duct where combustion products are flowing into the system (products similar to those resulting from exhaust in a spark ignited engine). The 2D-geometry then represents a co-flow of fresh air and burned gases containing UHC. The objectives of the simulations are the following:

- To test the model in a realistic configuration.

- To test the new developed CFD-FPI model over a broad range of initial conditions.

- Compare the influence of chemistry and physical mixing on reactivity under postoxidation conditions.

- To study the principles and properties of postoxidation in spark ignition engines induced by SAI.

8.1   Description of the co-flow simulation

The co-flow represents an axisymmetrical channel in which fresh air is injected parallel to an exhaust gases flow. After leaving the injection area, fresh air and exhaust gases mix. The exhaust gases contain combustion products and some unburned fuel and by mixing it with fresh air, the mixture becomes reactive. Once together, the exhaust gases and fresh air start to react and to auto-ignite after a certain ignition delay. Because the reactive mixture is highly diluted by combustion products, a fair temperature rise, like the one observed in a Diesel engine for example, is not observed. Mixture ignition in this particular case refers to the sudden production of combustion intermediates (CO, H_2...) and products (CO_2 and H_2O). The injected mass stream of fresh air $\dot{M}_{SAI}$ and the mass-stream of exhaust gases $\dot{M}_{BG}$ are kept constant, as well as the injection temperatures T_{SAI}, T_{BG}. The ignition delay time is linked with the distance between the gas injection and the location of main mixture reaction. Such co-flow allows observation of the mixing processes as well as the start of reaction in the flow region for various initial conditions.

8.1.1. Geometry and boundary conditions

Figure 8-1 shows a sketch of the simulated co-flow with its dimensions. Burned gases are injected along the radial outer channel named I_I. Fresh air is injected in the central channel I_{II} of length b along the axis of symmetry S. The computational area covers a length of 1 m and is limited by an outlet section O. In the radial direction, flows are limited by an adiabatic wall W.

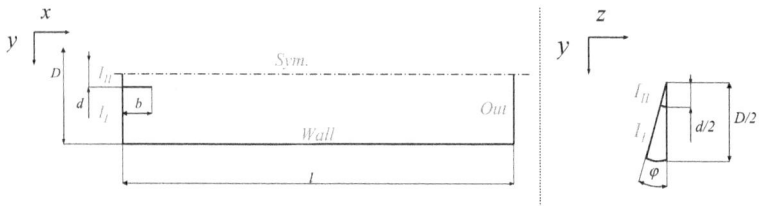

Figure 8-1   Co-flow geometry.

Table 29 lists the boundary conditions of the different computational sections defined in the co-flow model and Table 30 shows the dimensions defining the described computational area.

Table 29   Boundary conditions of the computational co-flow simulation.

Table 30   Dimensions of the computational co-flow simulation.

	Name	Boundary type
I_I	Inlet I	Constant mass stream
I_{II}	Inlet II	Constant mass stream
Out	Outlet	Constant pressure
Sym.	Symmetry	Symmetry
Wall	Wall	Adiabatic Wall

	Unit	size
l	[m]	1.0
D	[m]	0.07
d	[m]	0.001
B	[m]	0.01
φ	[rad]	0.1

The chosen diameters of the inlet section for fresh air injection I_{II} are of the same order of magnitude as injection diameters used for common SAI-systems.

8.1.2. Initialization of the co-flow simulations

CFD computations were initialized with an initial pressure P_0, a temperature T_0 and the initial gas composition over the complete computational area at the computation time $t_{comp} = 0$. The initial gas composition in the computational area at time $t_{comp} = 0$ corresponds exactly to the composition at which burned gases were injected in the inlet section I_I. For the performed computations H_2O and N_2 were considered as dilutants and the lower heating value of CO is represented within the fuel. The kinetic impact of CO_2 is not considered in the computations.

Mass-flow rates were estimated such that the reactants residence times in the computational area were around an order of seconds. The ratio of injected air and exhaust gas mass-flows for the *Reference-case* was fixed for an overall stoichiometry ϕ of 1. Mass-flows of injected air $\dot{M}_{SAI}$ and exhaust gases $\dot{M}_{BG}$ were kept constant for all performed computations in order to assure identical flow fields under all computed conditions. For the same reason, we did neither vary the temperature of the injected air T_{SAI}, nor the temperature of the burned gases T_{BG}. Such changes would lead to different gas densities and therefore to different injection velocities impacting the flow field.

The influence of temperature on mixture reactivity is generally well understood and for that reason, our parametric study was focused on the importance of exhaust gas composition on flow reactivity. We have thus defined a *Reference-Case* which was chosen so that the temperatures of a non-reactive perfect mixture of fresh air and exhaust gases is around 800 K. This corresponds to temperatures common for fresh air / exhaust-gas mixtures in the exhaust line before ignition (Kleemann, 2006). Such conditions were achieved for a temperature of injected fresh air T_{SAI} of 400 K and a temperature of the burned gases of 1000 K.

Table 31   Initial conditions of the CFD computations.

Case	$\dot{M}_{SAI}$	$\dot{M}_{BG}$	T_{SAI}	T_{BG}	P_0	y_{Fuel}	y_{NO}	y_{H2O}	ϕ
	[kg/h]	[kg/h]	[K]	[K]	[bar]	[kg/kg]	[kg/kg]	[kg/kg]	[-]
Reference-case	0.5	1.145	400	1000	1.0	0.03	0.0	0.07	1.0
Fuel_1	0.5	1.145	400	1000	1.0	0.049	0.0	0.07	0.5
Fuel_2	0.5	1.145	400	1000	1.0	0.011	0.0	0.07	2.2
NO_1	0.5	1.145	400	1000	1.0	0.03	5.0×10^{-5}	0.07	1.0
NO_2	0.5	1.145	400	1000	1.0	0.03	4.5×10^{-4}	0.07	1.0
P_1	0.5	1.145	400	1100	0.75	0.03	0.0	0.07	1.0
P_2	0.5	1.145	400	1100	1.5	0.03	0.0	0.07	1.0
Most Reactive-case	0.5	1.145	400	1000	1.5	0.049	5.0×10^{-5}	0.07	2.2

The reference pressure was the atmospheric pressure. The mass-flow rate of injected air was chosen very low (0.5kg/h) compared to mass flows imposed commonly to SAI systems (~20kg/h) (Kleeman, 2006). The purpose was to ensure long enough residence times for complete mixture reaction. In an engine, SAI is never injected axially, parallel to the out-flowing products. The flow velocity in a real system is therefore strongly impacted by a perpendicular air injection, for example. The mass flow of injected burned gases was defined as a function of the injected air mass-flow so that the overall equivalence ratio ϕ was equal to one. The fuel mass-fraction y_{Fuel} in the burned gases was chosen equal to 0.03 which corresponds to an engine global equivalence ratio ϕ_{engine} between 1.2 and 1.5. The mass fraction of NO was chosen equal to zero for the *Reference-Case*. We then varied the overall stoichiometry (by changing the content of initial fuel in the exhaust gases), the mass fraction of NO and finally, we tested the impact of pressure changes in the flow system. The initial conditions of all runs are listed in Table 31. The different initial conditions are named according to the corresponding variation compared to the *Reference-Case* (*Fuel_1, Fuel_2, NO_1...*). Fuel concentration y_{Fuel} was varied between 0.011 and 0.045 corresponding to changes in overall stoichiometry between $\phi \sim 0.5$ and $\phi \sim 2.2$. NO mass fractions were increased up to 4.5×10^{-4} (450 ppm). A more favourable case to fuel ignition was also studied.

8.2 Simulation results

Simulation results are first discussed for the *Reference-case* described in the previous section. Flow fields and species concentration profiles are shown. Then, parametric variations of fuel and NO concentrations in the exhaust and the impact of pressure were studied.

8.2.1. Reference-case

Figure 8-2 shows the computed fuel concentration field at the moment of ignition. Ignition was defined as the time at which, in at least one flow region, all oxygen or all fuel is consumed. In the *Reference-case* (Table 31), the ignition delay thus defined is 1.2 s.

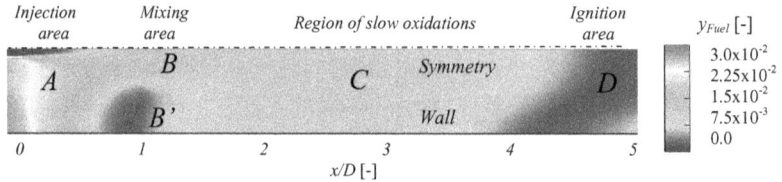

Figure 8-2   Fuel concentration field computed under the conditions corresponding to the Reference-case at the moment of ignition.

In Figure 8-2, the concentration field at the moment of ignition can be divided into four zones:

- Zone A is the region where fresh air and fuel contained in the exhaust gases first meet. This zone is characterized by areas of pure air (close to the axis) and pure exhaust gases where no mixing has yet occurred. In the axial direction, zone A is limited by the point at which fuel reaches the symmetry axis.

- Zone B corresponds to the mixing area between both streams. This region contains a back-flow vortex close to the wall (zone B'). The back-flow results from strong velocity gradients caused by the higher injection velocity of fresh air compared to exhaust gases. The recirculation leads to an increased mixture residence time. Some fuel conversion is expected to occur there.

- Zone C is characterized by total mixing over the complete radial section. In that area, fuel is partially oxidized, but neither fuel nor oxygen is completely consumed. Such behaviour is characteristic of slow oxidations.

- Zone D corresponds to the complete consumption of fuel or oxygen (depending on the stoichiometry). This region is where ignition occurs.

Our goal is to understand the ignition behaviour of the exhaust gases under turbulent flow conditions. We are aware that the chosen configuration is a very rough approach compared to flow conditions in the exhaust line of an IC-engine. However, it allows the study of kinetics effects in combination with physical mixing under postoxidation conditions. Thus, species evolutions in the computational area for different computational times were plotted. Species of interest are fuel and oxygen, as well as all tabulated combustion products (CO, CO_2, H_2O, and H).

Figure 8-3 shows the composition fields of fuel, O_2, CO, CO_2, H_2O and H at four different computational times t_{comp} (t_{comp} = 0.02 s, 0.76 s, 1.2 s and 2 s). The first computational time (t_{comp} = 0.02 s) shows the composition field at the start of injection, the second (t_{comp} = 0.76 s) represents the moment at which the mixture starts to react (cool flame), the third (t_{comp} = 1.2 s) corresponds to the ignition delay as defined previously and the last (t_{comp} = 2.0 s) is close to the steady-state solution.

shows for the same computational times (t_{comp} = 0.02 s, 0.76 s, 1.2 s and 2 s) flow velocitities, streamlines, the special distributions of temperature T, the equivalence ration ϕ and the reaction progress c.

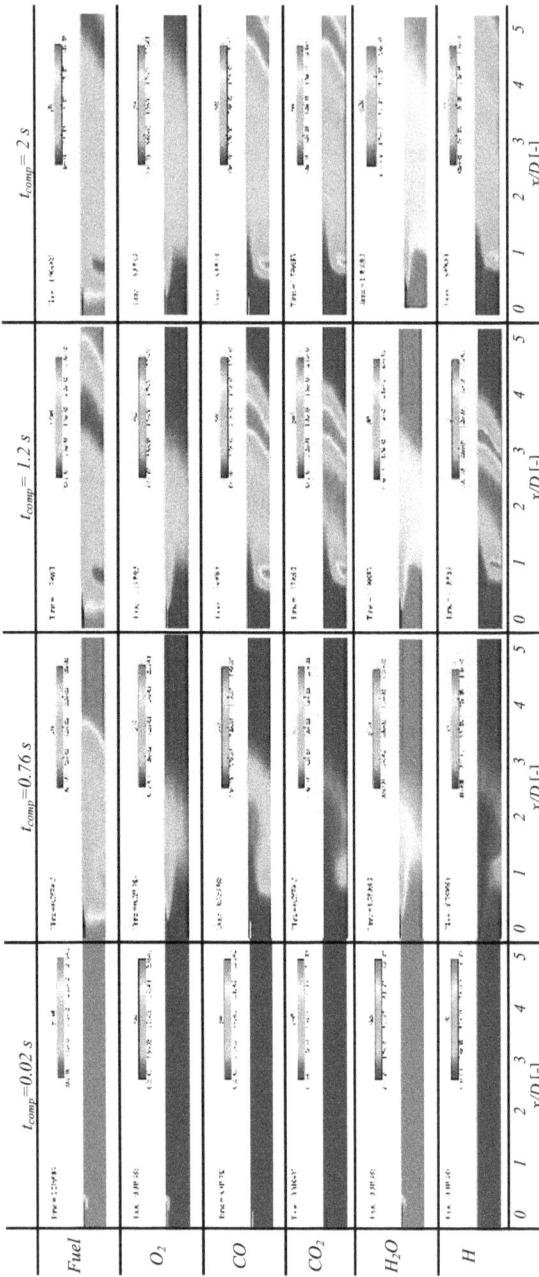

Figure 8-3   Species concentration fields computed under the conditions corresponding to the Reference-case at different computational times t_{comp}

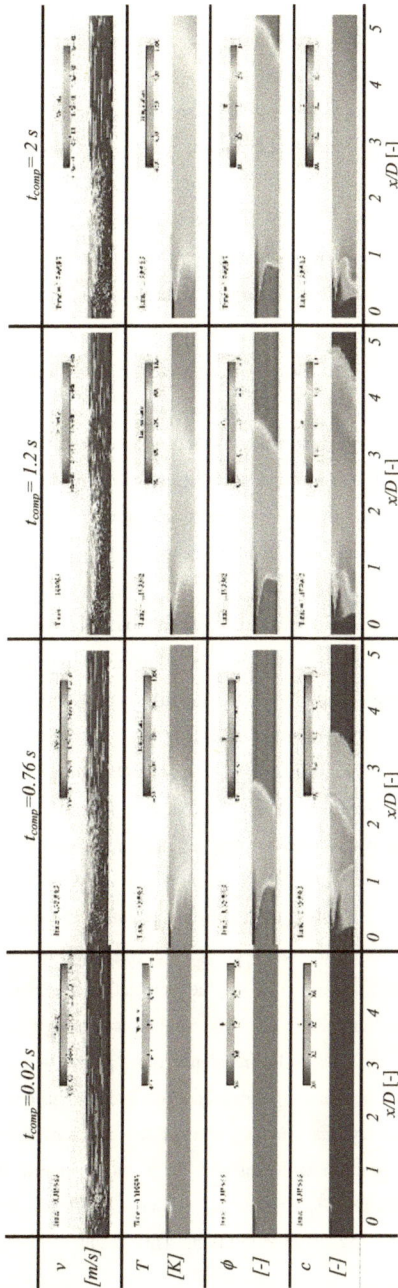

Figure 8-4 Physical properties of the computational field: Velocity, temperature, stoichiometry and reaction progress computed under the conditions corresponding to the Reference-case at different computational times t_{comp}.

At t_{comp} = 0.02 s, only exhaust gases are present in the system. This means that each computational cell contains a mixture of fuel, H_2O and N_2. The composition initialized at t_{comp} = 0 s corresponds exactly to that of injected exhaust gases in the inlet section I_I. Thus, at time t_{comp} = 0.12 s, besides fuel and H_2O, all other mass fractions are zero over almost the whole computational area. O_2 is present in the injection area close to inlet I_{II}, where the jet of fresh air starts to develop.

At computational time t_{comp} of 0.76 s, the jet of injected fresh air is fully developed and the fuel of the exhaust gases has mixed up with the oxygen present in the fresh air. The back-flow observed in zone B' in Figure 8-2 starts to develop as well as the mixing between exhaust gases and injected air. When the gases reach zone C (Figure 8-2), they are completely mixed. Thus, in the zone C, at time t_{comp} = 0.76 s, slow oxidations are observed leading to a slight consumption of fuel and oxygen and to the production of CO and CO_2. This is also true for H_2O but at this stage, the H_2O produced by oxidation is small compared to the amount injected in the burned gases as a dilutant. The production of H_2O is therefore not detectable in Figure-8-3. We have plotted as well the concentration profiles of tabulated atomic hydrogen H mass fractions because the presence of H is also a good indicator of zones with increased reactivity.

At time t_{comp} of 1.2 s, auto-ignition occurs. When the convected mixture has reached the autoignition zone D (Figure 8-2), the fuel and oxygen are completely consumed and transformed into CO and CO_2. Complete fuel and oxygen consumption is observed as well in the back-flow region B'. When the steady-state solution is close at t_{comp} = 2 s, one observes that the cool-flame region C expands and pushes the autoignition zone D downstream out of the computational domain so that it completely disappears. At this moment, all gases downstream of the mixing area are almost perfectly mixed and the global reactivity is low.

It is interesting to note that in this configuration, even if auto-ignition occurs at t_{comp} = 1.2 s, a steady-state reactive mixture is not established from there after. By pushing the combustion products out of the system, the mixture becomes highly homogeneous. This means that for the same global equivalence ratio close to one, the temperature becomes colder and the dilution is higher. Under the resulting conditions, auto-ignition is no longer possible inside the domain since the ignition delay exceeds the residence time. At t_{comp} = 1.2 s, the mixture ignites in a stratified, hotter and less diluted zone. One possible conclusion is that under postoxidation conditions, auto-ignition is favoured by mixture stratification where high equivalence ratios are observed. The ignition delay of rich mixtures is indeed shorter compared to leaner conditions (see Figure 5-1, chapter 5.1). Thus, non steady-state air injected into the exhaust gases might favour auto-ignition. This result confirms the observation of Kleemann, (2006) who tested different SAI setups and observed favoured postoxidation conditions for pulsating SAI compared to continuous SAI.

8.2.2. Parametric study

After studying the *Reference-case*, we performed parameter variations. We kept flow and mixing conditions (injection velocity) constant and varied exhaust gas composition as follows:

- Stoichiometry and degree of dilution (variations of the fuel concentration).
- NO mass fractions in the exhaust gases.
- Pressure in the system.

We have analyzed species mass fractions along the symmetry axis S of the computational area.

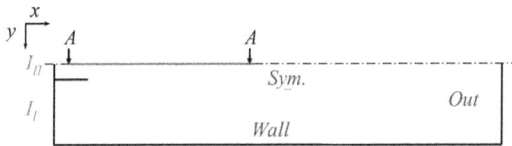

Figure 8-5   Section cut A-A in the computational area for the analysis of species mass fractions.

Figure 8-5 shows the location of the section cut where species mass fractions were analyzed and line A-A illustrates the distance over which mass fractions were plotted. All computed conditions of parameter variations are listed in Table 31. The analysis was performed for a computation time of 1.2 s, when main ignition was observed.

8.2.2.1. Variations of the global stoichiometry and dilution

The overall stoichiometry was varied by changing the fuel mass fraction in the exhaust gases and keeping the mass-flow of injected fresh air constant. Due to the architecture of the postoxidation FPI-model, a change of stoichiometry results automatically in a change of dilution ratio. As mentioned, the mass-fraction of initial CO_2 was set to 0 and thus equation [46] for computing N_2 mass fractions is reduced to equation [47]:

$$[47] \quad y_{N_2}^{init} = 1 - y_{Fuel}^{init} - y_{O_2}^{init} - y_{H_2O}^{init} - y_{NO}^{init}.$$

For that reason, the impact of dilution and stoichiometry could not be tested separately. Two variations were computed. In the first run, fuel mass fractions y_{Fuel} were reduced from 0.03 ($\phi = 1$, *Reference-case*) to 0.011 ($\phi \sim 0.5$, *Fuel_1*) and in the second run, they were increased up to 0.049 ($\phi \sim 2$, *Fuel_2*), (see Table 31). Figure 8-6 compares the mass-fractions of fuel, oxygen, CO, CO_2 and H_2O as a function of the non-dimensional distance x/D. x represents the axial distance from injection and D the diameter of the computational area.

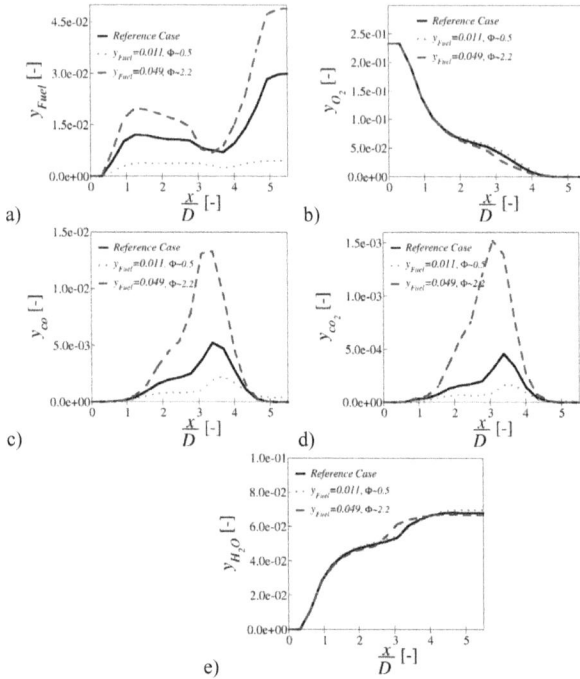

Figure 8-6   Mass fractions computed along the symmetry axis for a time t_{comp} of 1.2 s. Mass fractions of a) fuel, b) O_2, c) CO, d) CO_2 and e) H_2O are plotted as a function of the non dimensional axial distance x/D.

Close to the fresh air injection area at inlet I_{II} (x/D < 0.5), oxygen concentrations reach their maximum values while those of fuel and initial H_2O are zero. With increasing distance from the injection area, fuel is convected to the symmetry and thus, due to turbulent mixing, fuel and H_2O mass fractions increase constantly until x/D ~ 1. At x/D ~1, the mixture starts to react and a production of CO and CO_2 is observed. It reaches a maximum around x/D ~ 3.5. For higher values of x/D, CO and CO_2 mass fractions tend towards zero. This decrease of the combustion products is consistent with an increase of fuel mass fractions. However, oxygen concentrations remain constantly zero at x/D > 4 indicating that at the chosen computation time t_{comp} = 1.2 s the injected air has not been convected to the distance x/D > 5 and only gases present at t_{comp} = 0 s can be observed. With decreasing overall stoichiometry (y_{Fuel} =0.011, case Fuel_1) less fuel is convected towards the symmetry axis. A slight production of CO and CO_2 indicates reactivity at x/D >1. However, the fuel concentration remains almost constant over the complete distance up to the region where only initial gases are present at x/D >5. Thus, under poor conditions (and increased dilution ratio) postoxidation is strongly disfavoured compared to the *Reference-case*. The contrary is observed, when the overall stoichiometry is increased (y_{Fuel} =0.049, case Fuel_1). Fuel mass fraction increase to a local maximum at x/D ~1, similar to what is observed in the *Reference-case*. However, due to stronger reactivity compared to the *Reference-case* a pronounced decrease of fuel mass fractions is observed until x/D ~3.5. The increased production peaks of CO and CO_2 follow from the enhanced fuel reactivity under such conditions. These results reveal that an increased fuel concentration in the exhaust gases favours postoxidation due to increased overall stoichiometry and reduced dilution.

8.2.2.2. Variations of NO mass fractions

Two cases of different initial NO mass fraction were computed and compared to the *Reference-case*. In both cases the exhaust gases present at t_{comp} = 0 as well as the exhaust gases injected in the inlet I_1 (Figure 8-1) contained NO in different concentrations. Simulations were performed for NO mass fractions of y_{NO} = 5.0x10^{-5} (*NO_1*) and NO mass fractions y_{NO} = 4.5x10^{-4} (*NO_2*) (see Table 31). Figure 8-7 shows the computed mass-fractions of fuel, oxygen and the combustion products CO, CO_2 and H_2O in function of the non-dimensional distance x/D.

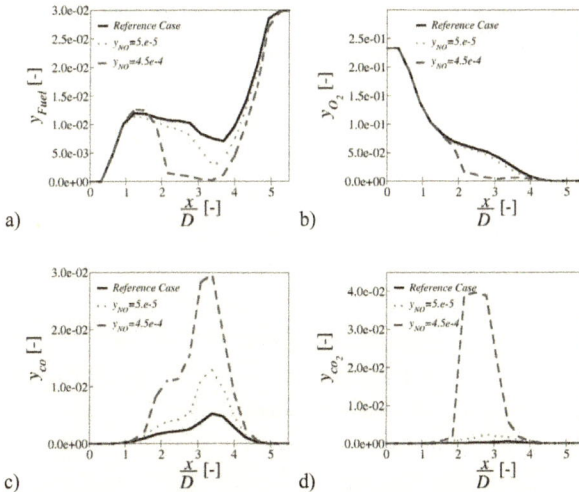

Figure 8-7   *Mass fractions computed along the axis of symmetry for a time t_{comp} of 1.2 s. Mass fractions of a) fuel, b) O_2, c) CO and d) CO_2 are plotted in function of the non dimensional axial distance x/D.*

When NO is present in the exhaust gases an impressive increase of reactivity is observed compared to the same computations in absence of NO (*Reference-case*). The fuel consumption observed between $x/D > 1$ and $x/D < 4$ increases with increasing NO concentrations. When NO is present in fractions $y_{NO} = 5.0 \times 10^{-5}$, the computed fuel concentration at $x/D \sim 3.5$ is almost half of that which was computed for the *Reference-case*. Such increased reactivity results in increased production peaks of CO and CO_2 around $x/D \sim 3.5$. However fuel is not completely consumed and thus oxidation is not complete.

Complete oxidation occurs, when NO is initialized at a mass fraction $y_{NO} = 4.5 \times 10^{-4}$. In presence of such NO concentrations fuel reactivity is strongly accelerated and between $x/D \sim 2$ and $x/D \sim 3$ all fuel and oxygen is consumed. This results in a strong increase of produced CO and a much stronger increase of CO_2 concentrations. The maximum CO_2 concentrations computed in the case of increased presence of NO ($y_{NO} = 4.5 \times 10^{-4}$, NO_2) overcome those computed for less NO present ($y_{NO} = 5.0 \times 10^{-5}$, NO_1) and the *Reference-case* by an order of magnitude. We thus conclude that the presence of NO promotes strongly postoxidation of exhaust gases.

8.2.2.3. Variations of the inital pressure

We varied as well the initial pressure in the computational area. Simulations were performed for a pressure of 0.75 bar (*P_1*) and of 1.5 bar (*P_2*) (see Table 31). Figure 8-8 displays the obtained mass-fractions of fuel, oxygen and the combustion products CO, CO_2 and H_2O and compares them to the *Reference-case*.

At a pressure of 0.75 bar, global reactivity decreases compared to the *Reference-case*. Fuel mass fractions in the area of slow oxidations (area *C*) between $x/D \sim 1$ and $x/D \sim 3$ are greater than those obtained for the *Reference-case*. Consequently the CO and CO_2 production is lower at a pressure of 0.75 bar than at atmospheric pressure. The peak of maximum CO_2 is shifted downstream. At increased pressures of 1.5 bar mixture reactivity is drastically favoured. The rate of fuel consumption in the area of slow oxidations between $x/D \sim 1$ and $x/D \sim 3$ is much greater than in the *Reference-case*. Consequently the CO and CO_2 production peaks are much more expressed than at atmospheric pressure. In the case of CO_2 maximum concentrations exceed at a pressure of 1.5 bar those obtained under atmospheric pressure by an order of magnitude. At 1.5 bar the peaks of maximum CO and CO_2 are shifted upstream.

We thus conclude that postoxidation of exhaust gases is sensitive to pressure fluctuations. Changes in pressure from atmospheric pressure to 1.5 bar may impact drastically fuel conversion and the production of CO and CO_2.

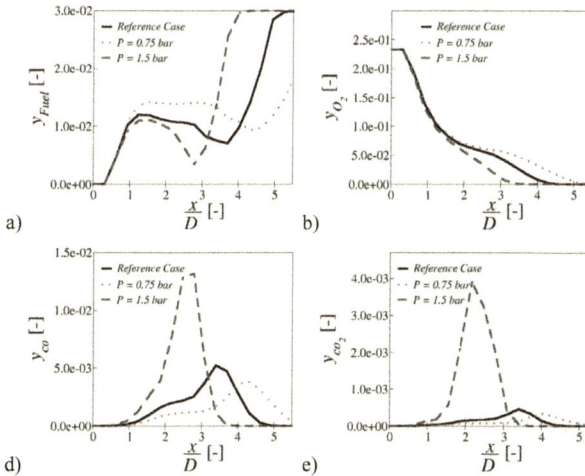

Figure 8-8   *Mass fractions computed along the symmetry axis for a time t_{comp} of 1.2 s. Mass fractions*
of a) fuel, b) O_2, c) CO, and d) CO_2 are plotted in function of the non dimensional axial
distance x/D.

8.2.2.4. Most reactive case

The parameter variations presented in the last sections revealed that postoxidation of exhaust gases is favoured:

- At rich stoichiometry

- In presence of NO

- At increased pressures

We were interested in autoignition under conditions most favourable to postoxidation of exhaust gases and performed a simulation under globally rich conditions ($y_{Fuel} = 0.049$), in presence of NO ($y_{NO} = 5.0 \times 10^{-5}$) and at elevated pressure ($P_0 = 1.5$ bar). The initial conditions the simulation run correspond to those of the *Most Reactive-case* in Table 31. The computed mass fractions of fuel, oxygen, CO, CO_2 and H_2O are shown in Figure 8-9 and compared to the case *Fuel_2* (rich conditions), *NO_1*, *P_2* (elevated pressure) and the *Reference-case*.

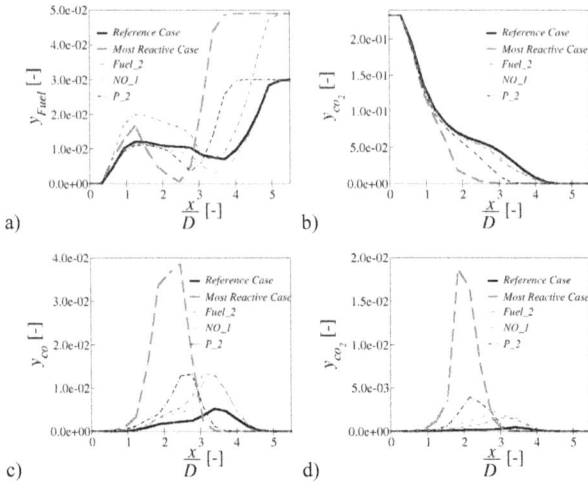

a) b)

c) d)

Figure 8-9   Mass fractions computed along the symmetry axis for a time t_{comp} of 1.2 s. Mass fractions of a) fuel, b) O_2, c) CO and d) CO_2 are plotted in function of the non dimensional axial distance x/D. Black full lines (—) correspond to results obtained for the Reference-case, green broken lines ( ) correspond to results obtained for the Most Reactive-case.

When cases *Fuel_2*, *NO_1* and *P_2* are combined to the *Most Reactive-case* a superposition of the effects of changes in stoichiometry and dilution ratio, presence of NO and increased pressure is observed. Increased stoichiomerty in presence of NO enhances drastically fuel convection in the region of slow oxidations (area *C*) (1 < x/D < 3). This intensifies when pressure exceeds the atmospheric pressure. Additionally the increase of pressure results in an upstream shift of the maximum CO and CO_2 production peaks.

shows the composition fields obtained for the *Most Reactive-case* of fuel, O_2, CO, CO_2, H_2O and H at different computational times t_{comp} (t_{comp} = 0.02 s, 0.76 s, 1.2 s and 2 s). Generally the maximum concentrations of CO and CO_2 exceed by far those obtained for the *Reference-case* (Figure-8-3) resulting from increased fuel concentrations in the injected exhaust gases. In contrast to what is observed for the *Reference-case*, in the *Most Reactive-case*, CO and CO_2 fields do not correspond to each other at increased residence times (t_{comp} = 1.2 s and 2 s). The CO_2 production peak occurs close to the symmetry axes next to the O_2-jet from injected. These areas locally lean or close to stoichiometry. Global stoichiometry of injected air and burned gases however are rich in the *Most Reactive-case*. Thus, reaction conditions are rich when gases are totally mixed downstream the CO_2-production peak and an increased CO formation is observed.

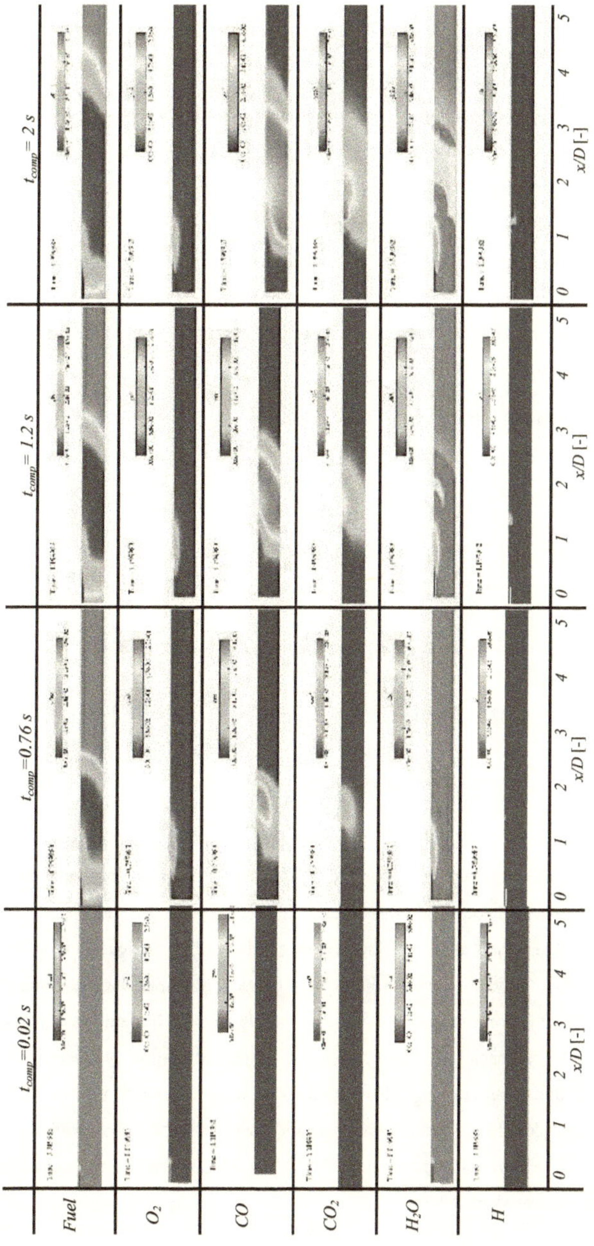

Figure 8-10 Species concentration fields computed under the conditions corresponding to the 'Most Reactive' case at different computational times t_{comp}.

8.3   Concluding remarks

The performed CFD-simulations allowed us to apply the new developed FPI-model on CFD-computations. The model was tested over a wide range of conditions. These tests were performed for physical conditions approaching those found in the exhaust line of an IC-engine equipped with a SAI-system. Mixing and chemical reactivity were simulated for a gaseous jet of fresh air which was injected into exhaust gases containing unburned fuel. The impact of changes in equivalence ratio, the effect of NO present in the mixture, as well as the influence of pressure on postoxidation was estimated. Computations led to the following conclusions:

- An instationary fresh air jet injected into exhaust gases favours auto-ignition compared to a stationary developed jet due to higher stoichiometric stratifications.

- An increase of fuel concentration in the exhaust gases favours postoxidation due to increased overall stoichiometry and reduced dilution.

- The presence of NO promotes strongly postoxidation of exhaust gases.

- Postoxidation of exhaust gases is sensitive to pressure fluctuations. Slight variations in pressure may change drastically reactivity. Postoxidation is favoured at increased pressures.

The observed ignition delay times for that computational configuration range in seconds as an order of magnitude. However, observed ignition delay times are much shorter when exhaust gases are postoxidized inside the exhaust lane of an IC-engine equipped with SAI (Kleemann, 2006). This result as well as the performed study of postoxidation ignition limits shown in section 5.3, indicates that the back flow of exhaust gases during exhaust valve opening into the combustion chamber plays a determining role for burned gas postoxidation induced by SAI. The successful computations on the simplified co-flow geometry as well as the fact that a full understanding of postoxidation induced by SAI needs a modelling of the combustion chamber coupled to the exhaust manifold encourage to implement the proposed FPI-model into a 3D-engine configuration. For such 3-D engine computations, a complete engine combustion chamber should be modelled as well as the engine exhaust line equipped with SAI.

Chapitre 9

Conclusions

Within the present work, a detailed kinetic reaction model for the oxidation of PRF/toluene mixtures sensitized by NO has been developed. This model was validated over a wide range of conditions covering postoxidation characteristics in IC-engines. The kinetic development was the basis for further studies giving detailed insight on the formation of C_1-C_2 species from larger hydrocarbons. The kinetic impact of dilution gases on hydrocarbon oxidation kinetics and ignition limits of exhaust gases was studied.

The new developed detailed reaction mechanism was coupled with a CFD-code using a FPI-tabulation methodology. The developed FPI-model was conceived for postoxidation applications and describes the oxidation of unburned hydrocarbons in turbulent flowing low temperature, low pressure and highly diluted exhaust gases. The model was tested in a configuration representing air injection into exhaust gases, under conditions typical of an IC-engine exhaust line equipped with SAI. In the following sections, conclusions that may be drawn from the present work performed on chemical kinetic and FPI modelling are discussed.

9.1   Kinetic development

The detailed kinetic reaction mechanism developed for PRF/toluene blends was successfully validated against different experimental applications over a wide range of thermochemical conditions. These validations showed that the kinetic model reproduces correctly the chemistry under postoxidation conditions. Sensitivity and flux analysis gave detailed insight into governing reactions and their interactions.

In our study on the formation of C_1-C_2 species, we observed that the actual model overestimates the production of these species at temperatures below 800 K. This lead to a revision of the secondary sub-set of the mechanism produced by EXGAS, an automatic mechanism generator, and we could drastically improve the prediction of C_1-C_2 species formation at low temperatures. Such improvement enlarges the potential application of the mechanism for CFD-combustion models. C_1-C_2 species are precursors of soot formation and their correct prediction is fundamental for CFD-models describing the formation of soot during hydrocarbon oxidation. The proposed improvements of the C_1-C_2 EXGAS sub-set were however not included in the final PRF/toluene mechanism used in the CFD modeling part. The study of the secondary mechanism has in fact excluded iso-octane, a necessary component of the PRF/toluene gasoline surrogate necessary for the internal combustion engine CFD studies.

A study on the influence of residual gases on hydrocarbon oxidation was then performed. It revealed that under postoxidation conditions, major combsution products CO_2 and H_2O may exhibit a kinetic impact when present in the dilutant. CO also has a remarkable effect on hydrocarbon oxidation. We showed that collision efficiencies of different collision partners play a major role for low temperature H_2O_2 dissociation kinetics. Especially, H_2O as a collision partner may impact H_2O_2 dissociation. Thus, for a convenient modelling of postoxidation kinetics, the kinetic impact of dilutant gases on hydrocarbon oxidation cannot be neglected.

We have also observed that compared to n-heptane, iso-octane shows quite particular oxidation characteristics when mixed with toluene. These are not captured accurately neither by the actual reaction model nor by such other models known from the literature. We have tested alternative reaction pathways including an additional propagation step by the formation of cyclic ethers form iso-octylperoxy radicals and by defining additional cross reactions for cyclic ether interactions with toluene. First tests (not shown in this work) revealed that promising results can be obtained and can improve the understanding of iso-octane oxidation and its co-oxidation with toluene.

9.2 FPI-modelling

FPI- tabulation methods were used to couple the developed detailed reaction mechanism with the CFD-code *IFP-C3D*. A FPI-model was designed for CFD-computations of the postoxidation of unburned hydrocarbons in IC-engine exhaust gases. The major challenge was the identification and implementation of initial thermodynamic conditions $(P_0(t),\ T_0(t),\ \phi_0(t)...)$ governing mixture reactivity. Thermodynamic conditions in the exhaust line differ strongly from those found during IC-engine combustion inside a combustion chamber. Thus, the architecture of the developed FPI-model does not correspond to that of a FPI-table designed for engine computations. When postoxidation kinetics are tabulated, the number of table dimensions, and table size increases drastically because of the integration of chemical species as table dimensions and must therefore be optimized. The proposed new FPI-model reduces table dimensions and size by correlating the dilutant composition to the unburned fuel present in the exhaust gases. The amount of unburned fuel in the engine exhaust gases is estimated by closure of energy balances and the composition of the diluting gases is correlated to the stoichiometry at which the fresh gases were burned inside the IC-engine. The application of the FPI-postoxidation model was illustrated by simulating an IC-engine exhaust gas postoxidation induced by SAI. Unburned hydrocarbons are represented by the initial fuel. The next development step should be the implementation of information about fuel desorbed from crevices and walls into the FPI-table. Therefore fuel desorbed from crevices could be represented by the initial fuel, whereas unburned hydrocarbons could be substituted by CO and toluene, which showed similar ignition delay times compared to those computed for realistic exhaust gas compositions.

As an illustration of the CFD postoxydation model application, we have modelled a co-flow geometry, where fresh air was injected into exhaust gases containing unburned initial fuel. Simulated thermodynamic and flow conditions correspond to those typical of an IC-engine exhaust line equipped with SAI. The computations showed that reactivity of postoxidized turbulent exhaust gases are sensitive to:

- Equivalence ratio stratification.

- The presence of small amounts of NO.

- Pressure fluctuations.

These computations further showed that our approach is fit for CFD-computations including complex chemistry valid with highly diluted gases in low and moderate temperatures. Thus, the range of potential model applications includes engine postoxidation computations as well as highly diluted combustion modes like mild combustion applications for example.

9.3 Perspectives

This work covered two areas of development, the kinetic modelling and the development of a new FPI-tabulation approach of detailed chemistry under postoxidation conditions for complex CFD computations. Future work continuing the kinetic development made here should focus on:

- A better understanding of the iso-octane oxidation kinetics and its co-oxidations with aromatic compounds.

- The enhanced secondary mechanism developed for n-heptane/toluene mixtures should coupled with iso-octane and implemented in the mechanism for PRF/Toluene/NO blends

- Kinetics of H_2O_2-dissociation should be revised. Uncertainties on the collision efficiencies of collision partners exist. Further experimental as well as numerical investigations on the topic might be helpful to improve kinetic modelling. In the present reaction mechanism, the user has to change rate coefficients of the H_2O_2-dissociation for the correct prediction of ignition delay times at high pressures ($P > 40$ bar). This is certainly due to the above studied uncertainties of collision efficiencies for 3rd-Body collision partners.

Concerning future developments of the present FPI-model, the following deveopments should be performed:

- Fuel emitted from crevices should be considered in the model. Partially oxidized hydrocarbons issued from incomplete combustion might be substituted by a mixture of toluene and methane, while the fuel emitted from crevices could be substituted by the initial PRF/toluene mixture. Such an approach could allow as well the correlation of CO and CO_2 with the mass fraction of partially oxidized hydrocarbons, as it was proposed for H_2O.

- The potential application of the FPI-model should be extended to various fuel types like oxygenated fuels (alcohols, esters) or natural gas. Therefore, the impact of different fuel types on exhaust gas reactivity should be tested in a similar way as we have shown in the present work for NO. Look-up tables for adequate fuel surrogates should be generated. 'Key'-species of such oxygenated fuels must then be identified and corresponding reaction mechanisms developed. Thus, for each fuel type of interest, a separate FPI-table should be generated and coupled to CFD-computations.

Liste of publications and communications

Publications in international reviews

ANDERLOHR J.M., PIPEREL A., PIRES DA CRUZ A., BOUNACEUR R.,
BATTIN-LECLERC F., DAGAUT P., MONTAGNE X. (2009).
Influence of EGR Compounds on the Oxidation of an HCCI-Diesel Surrogate.
Proc. Combust. Inst., 32, 2851-2859.

ANDERLOHR J.M.,D BOUNACEUR R., PIRES D.A. CRUZ A., BATTIN-LECLERC F.
(2009).
Modelling of Autoignition and NO Sensitization for the Oxidation of IC-Engine Surrogate Fuels.
Combust. Flame, 156, 505-521.

ANDERLOHR J.M., PIRES DA CRUZ A, BOUNACEUR R, BATTIN-LECLERC F.,
(2010).
Thermal and kinetic Impact of CO, CO_2 and H_2O on the Postoxidation of IC-Engine Exhaust Gases.
Combst. Sci. Technol., 182, 30-59.

Communications on international conferences:

ANDERLOHR J., PIRES DA CRUZ A., BOUNACEUR. R., BATTIN LECLERC F.
Modeling of NO sensitization of IC engines surrogate fuels auto-ignition and combustion.
Communication orale au *"21st International Colloquium on the Dynamics of Explosions and Reactive Systems"*, Poitiers (FRANCE), 23-27 juillet 2007 : article (4 pages) inclus sur le CD-ROM du congrès (2007).

PIRES DA CRUZ A., PERA C., ANDERLOHR J., BOUNACEUR R.,
BATTIN-LECLERC F.
A complex chemical kinetic mechanism for the oxidation of gasoline surrogate fuels: n-heptane, iso-octane and toluene – Mechanism development and validation.
Communication par affiche au *"Third European Combustion Meeting ECM 2007'"*, Chania (GRECE), 11-13 avril 2007 : article (6 pages) inclus sur le CD-ROM du congrès (2007).

ANDERLOHR J.M., PIPEREL A., PIRES DA CRUZ A., BOUNACEUR R., BATTIN-LECLERC F., DAGAUT P., MONTAGNE X.
Influence of EGR Compounds on the oxidation of an HCCI-Diesel surrogate.
Communication orale au *"32nd International Symposium on Combustion"*, Montréal (CANADA), 3-9 Août 2008 : article correspondant à la publication P_2, (2008).

Bibliography

ALZUETA M.U., GLARBORG P., DAM-JOHANSEN K. (1997).
Low Temperature Interactions Between Hydrocarbons and Nitric Oxide: An experimental Study.
Combust. Flame, 109, 25-36.

ALZUETA M.U., BIBAO R., FINESTRA M. (2001).
Methanol Oxidation and Its Interactions with Nitric Oxide.
Energy and Fuels, 15, 724-729.

ANDERLOHR J.M., BOUNACEUR R., PIRES D.A. CRUZ A., BATTIN-LECLERC F. (2009a).
Modelling of Autoignition and NO Sensitization for the Oxidation of IC-Engine Surrogate Fuels.
Combust. Flame, 156, 505-521.

ANDERLOHR J.M., PIPEREL A., PIRES DA CRUZ A., BOUNACEUR R., BATTIN-LECLERC F., DAGAUT P., MONTAGNE X. (2009b).
Influence of EGR Compounds on the oxidation of an HCCI-Diesel surrogate.
Proc. Combust. Inst., 32, 2851-2859.

ANDRAE J., JOHANSSON D., BJÖRNBOM P., RISBERG P., KALGHATGI G. (2005).
Co-oxidation in the auto-ignition of primary reference fuels and n-heptane/toluene blends.
Combust. Flame, 140, 267-286.

ANDRAE J.C.G., BJÖRNBOM P., CRACKNELL R.F. ET KALGHATGI G.T. (2007).
Autoigniton of toluene fuels at high pressures modeled with detailed chemical kinetics.
Combust. Flame, 149, 2-24.

AMERICAN SOCIETY FOR TESTING AND MATERIALS (1993).
Annual Book of ASTM Standards. Test methods for rating motor, diesel and aviation fuels (05.04).

AMSDEN A.A., ROURKE P.J.O., BUTLER T.D. (1989).
KIVAII: A Computer Program for Chemically Reactive Flows with Sprays.
Report No. LA-11560-MS.UC-96, Los Alamos National Laboratory.

ATKINSON R., BAULCH D.L., COX R.A., HAMPSON R.F., KERR J.A., TROE J. (1992).
Evaluated kinetic and photochemical data for atmospheric chemistry supplement –IV – IUPAC subcommittee on gas kinetic data evaluation for atmospheric chemistry.
J. Phys. Chem. Ref. Data 21: p 1125.

ATKINSON R., BAULCH D.L., COX R.A., CROWLEY J.N., HAMPSON R.F., HYNES R.G., JENKIN M.E., ROSSI M.J., TROE J. (2004)
Evaluated kinetic and photochemical data for atmospheric chemistry: Volume I – gas phase reactions of Ox, HOx, NOx and SOx species.
Atmps. Chem. Phys., 4, 1461-1738.

ATKINSON R., BAULCH D.L., COX R.A., CROWLEY J.N., HAMPSON R.F., HYNES R.G., JENKINS M.E., ROSSI M.J., TROE J. (2005).
Evaluated kinetic and photochemical data for atmospheric chemistry: Volume II – reactions of organic species.
Atmos. Chem. Phys. Discuss., 5, 6295-7168.

BALDAVIN R.R., DEAN C.E., WALKER R.W. (1986).
Relative rate study of the addition of HO_2 radicals to C_2H_4 and C_3H_6.
Chem. Soc. Faraday Trans., 2 (9), 1445-1455.

BATTIN-LECLERC F., FOURNET R., GLAUDE P.A., JUDENHERC B., WARTH V.
CÔME G.M., SCACCHI G. (2000).
Modeling of the gas-phase oxidation of n-decane from 550-1600K.
Proc. Combust. Inst., 28, 1597-1605.

BATTIN-LECLERC F. (2008).
Detailed chemical kinetic models for the low-temperature combustion of hydrocarbons with application to gasoline and diesel fuel surrogates.
Prog. Energ. Combust. Sci., 34, 440-498.

BARBÉ P., BATTIN-LECLERC F., CÔME G.M. (1995).
Experimental and modelling study of methane and ethane oxidation between 773K and 1573K.
J. Chim. Phys., 92, 1666.

BAULCH D.L., COBOS C.J., COX R.A., ESSER C., FRANK P. (1992).
Evaluated Kinetic Data for Combustion Modeling.
J. Phys. Chem. Ref. Data, 21, 411.

BAULCH D.L., COBOS C.J., COX R.A., FRANK P., HAYMAN G.D., JUST T., KERR J.A., MURRELLS T.P., PILLING M.J., TROE J., WALKER R.W., WARNATZ J. (1994).
Evaluated Kinetic Data for Combustion Modeling.
J. Phys. Chem. Ref. Data, 23, 847.

BAULCH D.L., BOWMAN C.T., COBOS C.J. (2005).
Evaluated Kinetic Data for Combustion Modeling: Supplement II.
J. Phys. Chem. Ref. Data, 34, 757.

BENSON S.W. (1976)
Thermochemical kinetics.
2nd ed. John Wiley, New York, 1976.

BIET J., HAKKA M.H., WARTH V., GLAUDE P.A., BATTIN-LECLERC F. (2008).
Experimental and Modeling Study of the Low-Temperature Oxidation of Large Alkanes.
Energy & Fuels, 22 (4), 2258-2269.

BLOCH-MICHEL V. (1995).
Logiciel d'estimation de paramètres cinétiques de processus élémentaires en phase gazeuse.
Thèse de l'Institut National Polytechnique de Lorraine, Nancy.

BOUNACEUR R., D COSTA R., FOURNET R., BILLAUD F., BATTIN-LECLERC F. (2005).
Experimental and Modeling Study of the Oxidation of Toluene.
Int. J. Chem. Kinet., 37-1, 25-49.

BOWMANN C.T,. HANSON R.K., DAVIDSON D.F., GARDINER W.C. JR. ET AL. (1997).
GRI-MECH 2.11.
http://www. me.berkeley.edu/gri_mech.

BRABBS T.A., LEZBERT E., BITTKER A. (1987).
Hydrogen Oxidation Mechanism with applications to (1) the Chaperon Efficiency of Carbon Dioxide and (2) vitiated Air Testing.
NASA Technical Memorandum 100186.

BREZINSKY K., LINTZINGER T., GLASSMAN I. (1984).
The high temperature oxidation of the methyl side chain toluene.
Int. J. Chem. Kinet., 16, 1053–1074.

BROMLY J.H., BARNES F.J., MANDYCZEWSKY R., EDWARDS T.J., HAYNES B.S. (1992).
An Experimental Investigation of the Mutually Sensitized Oxidation of Nitric Oxide and n-Butane.
Proc. Combust. Inst., 24, 899-907.

BROMLY J.H., BARNES F.J., MURIS S., YOU X., HAYNES B.S. (1996).
Kinetic and Thermodynamic Sensitivity Analysis of the NO-sensitized Oxidation of Methane.
Combust. Sci. Tech., 115, 259-296.

BRYUKOV M.G., KACHANOV A.A., TOMONNEN R., SEETULA J., VANDOREN J., SARKISOV O.M. (1993).
Chem. Phys. Letters, 208, 392-398.

BUDA F., BOUNACEUR R., WARTH V., GLAUDE P.A., FOURNET R., BATTIN-LECLERC F. (2005).
Progress toward a unified detailed kinetic model for the autoignition of alkanes from C_4 to C_{10} between 600 and 1200 K.
Combust. Flame, 142, 170-186.

BUDA F. (2006).
Mécanismes cinétiques pour l'amélioration de la sécurité des procédés d'oxydation des hydrocarbures.
Thèse de l'Institut National Polytechnique de Lorraine, Nancy.

BURCAT A., RUSCIC B. (2005).
Third Millennium Ideal Gas and Condensed Phase Thermochemical Database for Combustion with updates from Active Thermochemical Tables.
ANL-05/20 and TAE 960 Technion-IIT, Aerospace Engineering, and Argonne National Laboratory, Chemistry Division

CALLAHAN C.V., DRYER F.L., HELD T.J., MINETTI R., RIBAUCOUR M., SOCHET L.R., FARAVELLI T., GAFFURI P., RANZI E. (1996).
Experimental Data and Kinetic Modeling of Primary Reference Fuel Mixtures
Proc. Combust. Inst., 26, 739-746.

CHAN W.T., HECK S.M., PRITCHARD H.O. (2001).
Reaction of nitrogen dioxide with hydrocarbons and its influence on spontaneous ignition. A computational study.
Phys. Chem. Phys., 3, 56-62.

CHAKIR A., BELLIMAM M., BOETTNER J.C, CATHONNET M. (1992).
Kinetic study of n-heptane oxidation.
Int. J. Chem. Kinet., 24, 385-410.

CHAOS M., ZHAO Z., KAZAKOV A., GOKULAKRISNAN P., ANGIOLETTI M., DRYER F.L. (2007).
A PRF+toluene surrogate fuel model for simulating gasoline kinetics.
In: Proceedings of the fifth US Combustion Institute meeting, San Diego.

CHEVALIER C., PITZ W.J., WARNATZ J., WESTBROOK C.K., MELENK H. (1992).
Hydrocarbon Ignition: Automatic Generation of Reaction Mechanisms and Applications to Modeling of Engine Knock.
Proc. Combust. Inst., 24, 93-101.

CHOI Y.M., LIN M.C. (2005).
Kinetics and mechanisms for reactions of HNO with CH3 and C6H5 studied by quantum-chemical and statistical-theory calculations.
Int. J. Chem. Kinet., 37, 261-274.

CIEZKI H.K., ADOMEIT G. (1993).
Shock-tube investigation of self-ignition of n-heptane - Air mixtures under engine relevant conditions.
Combust. Flame, 93, 421-433.

CÔME G.M., WARTH V., GLAUDE P.A., FOURNET R., BATTIN-LECLERC F., SCACCHI G. (1996).
Computer-aided design of gas-phase oxidation mechanisms: Application to the modeling of n-heptane and iso-octane oxidation.
Proc. Combust. Inst., 26, 755.

COLIN O., BENKENIDA. A., ANGELBERGER C. (2003)
3D Modeling of Mixing, ignition and combustion phenomena in highly stratified gasoline engine.
Oil & Gas Science and Technology, 58-1, 47-52.

COLIN O. AND BENKENIDA A. (2004).
The 3-Zones Extended Coherent Flame Model (ECFM3Z) for computing premixed/diffusion combustion.
Oil & Gas Science and Technology - Rev. IFP, 59-6.

CURRAN H.J., GAFFURI P., PITZ W.J., WESTBROOK C.K. (1998).
A Comprehensive Modeling Study of n-Heptane Oxidation.
Combust. Flame, 114, 149-177.

CURRAN H.J., GAFFURI P., PITZ W.J., WESTBROOK C.K. (2002).
A Comprehensive Modeling Study of iso-Octane Oxidation.
Combust. Flame 129, 253-280.

CUOCI A., FRASSOLDATI A., BUZZI FERRARIS G., FARAVELLI T., RANZI E. (2007).
The ignition, combustion and flame structure of carbon monoxide/hydrogen mixtures. Note 2: Fluid dynamics and kinetic aspects of syngas combustion.
Int. J. Hydrogen Energ., 32, 15 3486-3500.

DA COSTA I., FOURNET R., BILLAUD F., BATTIN-LECLERC F. (2003).
Experimental and modeling study of the oxidation of benzene.
Int. J. Chem. Kinet., 35, 503-524.

DAGAUT P., CATHONNET M., ROUAN J.P., FOULATIER R., QUILGARS A., BOETTNER J.C., GAILLARD F., JAMES H. (1986).
A jet-stirred reactor for kinetic studies of homogeneous gas-phase reactions at pressures up to ten atmospheres (≈1 MPa).
J. Phys. E: Sci. Instrum., 19, 207-209.

DAGAUT P., REUILLON M., CATHONNET M. (1994).
High pressure oxidation of liquid fuel from low to high temperature. 2. Mixtures of n-heptane and iso-octane.
Combust. Sci. Technol., 103, 315-336.

DAGAUT P., REUILLON M., CATHONNET M. (1995).
Experimental study of the oxidation of n-heptane in a jet stirred reactor from low to high temperature and pressures up to 40 atm.
Combust. Flame, 101, 132-140.

DAGAUT P., LECOMTE F., CHEVAILLER S.. CATHONNET M. (1999).
Experimental and Kinetic Modeling of Nitric Oxide Reduction by Acetylene in an Atmospheric Pressure Jet-Stirred Reactor.
Fuel, 78, 1245-1252.

DAGAUT P., LECOMTE F., CHEVALLIER S., CATHONNET M. (1999).
The reduction of NO by ethylene in a jet-stirred reactor at 1 atm: experimental and kinetic modelling.
Combust. Flame, 119, 494-504.

DAGAUT P., LUCHE J., CATHONNET M. (2000).
Experimental and kinetic modeling of the reduction of NO by propene at 1 atm.
Combust. Flame, 121, 651-661.

DAGAUT P., LUCHE J., CATHONNET M. (2001).
Reduction of NO by propane in a JSR at 1 atm: experimental and kinetic modeling.
Fuel, 80, 979-986.

DAGAUT P., PENGLOAN G., RISTORI A. (2002).
Oxidation, ignition and combustion of toluene: Experimental and detailed chemical kinetic modelling.
Phys. Chem. Chem. Phys, 4, 1846-1854.

DAGAUT P., NICOLLE A. (2005).
Experimental study and detailed kinetic modeling of the effect of exhaust gas on fuel combustion: mutual sensitization of the oxidation of nitric oxide and methane over extended temperature and pressure ranges.
Combust. Flame, 140, 161-171.

DAVIDSON D.F., GAUTHIER G.M., HANSON R.K. (2005).
Shock tube measurements of iso-octane/air and toluene/air at high pressure.
Proc. Combust. Inst., 30, 1175–1182.

DAVIS S.G., LAW C.K. (1998).
Laminar flame speed and oxidation kinetics of iso-octane/air and n-heptane/air flames.
Proc. Combust. Inst., 27. 521-527.

DAVIS S.G., JOSHI A.V., WANG H., EGOLFOPOULOS F. (2005).
An optimized kinetic model of H2/CO combustion.
Proc. Combust. Inst., 30, 1283.

DEAN A.M., BOZZELLI J.W. (1997).
Combustion Chemistry
Gardiner, W. Jr. Ed. Springer-Verlag: New York, 1997.

DIAU E.W., HALBGEWACHS M.J., SMITH A.R., LIN M.C. (1995).
Thermal Reduction of NO by H_2: Kinetic Measurement and Computer Modeling of the HNO+NO Reaction.
Int. J. Chem. Kinet., 27, 867-881.

DIXON-LEWIS G., WILLIAMS D.J. (1977).
Comprehensive chemical kinetics
C. H. Branford and C.F. H. Tipper, eds. Vol. 17, Elsevier, Amsterdam, 1-248.

DJURISIC Z.M., JOSHI A.V., WANG H. (2001).
Detailed kinetic modelling of benzene and toluene combustion.
In: Proceedings of the second joint meeting of the US sections of the Combustion Institute, Oakland, 2001.

DUBREUIL A., FOUCHER F, MOUNAIM-ROUSSELLE C., DAYMA G., DAGAUT P. (2006).
HCCI combustion: Effect of NO in EGR.
Proc. Combst. Inst., 31, 2879-2886.

DUCLOS J.M., ZOLVER M. (1998).
3D Modeling of Intake, Injection and Combustion in a DI-SI Engine under Homogeneous and Stratified Operating Conditions.
International Symposium COMODIA.

EMBOUAZZA M. (2005).
Etude de l'Auto-Allumage par réduction des Schémas Cinétiques Chimiques. Application à la combustion homogène Diesel.
Thèse de l'Ecole Centrale Paris.

EMDEE J.L., BREZINSKY K., GLASSMAN I. (1992).
A kinetic model for the oxidation of toluene near 1200 K.
J. Phys. Chem., 96, 2151-2161.

ERRICO G., ONORATI A. (2002).
Modelling the Pollutant Emissions from a SI Engine.
SAE, 2002-01-0006.

ERRICO G., ONORATI A. (2003).
Secondary Air Injection in the Exhaust After Treatment System of S.I. Engines: 1D Fluid Dynamic Modeling and Experimental Investigation.
SAE, 2003-01_03667.

FARREL J.T., JOHNSTON R.J., ANDROULAKIS I.P. (2004).
Molecular Structure Effects On Laminar Burning Velocities At Elevated Temperature And Pressure.
SAE 2004-01-2936.

FIEWEGER K., BLUMENTHAL R., ADOMEIT G. (1997).
Self-ignition of S.I. engine model fuels: A shock tube investigation at high pressure.
Combust. Flame, 109, 599-619.

FIORINA B., BARON R., GICQUEL O., THEVENIN D., CARPENTIER S. DARABIHA N. (2003).
Modelling no-adiabatic partially premixed flames using flame-prolongation of ILDM.
Combust. Theory Modelling, 7(3), 449-470.

FLYNN P.F., HUNTER G.L., FARRELL L.A., DURRETT R.P., AKINYEMI O., WESTBROOK C.K., PITZ W.J. (2000).
The Inevitability of Engine-Out NOx Emissions from Spark-Ignition and Diesel Engines Chemical Kinetics of Hydrocarbon Ignition in Practical Combustion Systems.
Proceedings of the Combustion Institute, 28, 1211-1218.

FRASSOLDATI A., FARAVELLI T., RANZI E. (2003a).
Kinetic modeling of the interactions between NO and hydrocarbons in the oxidation of hydrocarbons at low temperatures.
Combust. Flame, 132, 188-207.

FRASSOLDATI A., FARAVELLI T., RANZI E. (2003b).
Kinetic modeling of the interactions between NO and hydrocarbons at high temperature
Combust. Flame ,135, 97-112.

FRASSOLDATI A., FARAVELLI T., RANZI E. (2007).
The ignition, combustion and flame structure of carbon monoxide/hydrogen mixtures. Note 1: Detailed kinetic modeling of syngas combustion also in presence of nitrogen compounds.
Int. J. Hydrogen Energ., 32-15, 3471-3485

GARDINER W.C., OLSON D.B. (1980).
Chemical kinetics of high temperature combustion.
Ann. Rev. Phys. Chem., 31, 377.

GAUTHIER B.M., DAVIDSON D.F., HANSON R.K. (2004).
Shock tube determination of ignition delay times in full-blend and surrogate fuel mixtures.
Combust. Flame, 139, 300-311.

GICQUEL O.,DARAHIBA N., THEVENIN D. (2000).
Laminar premixed hydrogen/air count rflow flame simulations using flame prolongation of ildm with differential diffusion.
Proc. Combust. Inst., 28, 1901-1908.

GLÄNZER K., TROE J. (1972).
Thermische Zerfallsreaktionen von Nitroverbindungen I: Dissoziation von Nitromethan.
Helvetica Chimica Acta, 55, 2884–2893.

GLARBORG P., ALZUETA M.U., DAM-JOHANSEN K., MILLER J.A. (1998).
Kinetic Modeling of Hydrocarbon/Nitric Oxide Interactions in a Flow Reactor.
Combust. Flame, 115, 1-27.

GLARBORG P., BENDTSEN A.B., MILLER J.A. (1999).
Nitromethane Dissociation : Implications for the CH_3+NO_2 Reaction.
Int. J. Chem. Kinet., 31, 591-602.

GLARBORG P., BENDTSEN L.B. (2008).
Chemical Effects of a High CO_2 Concentration in Oxy-Fuel Combustion of Methane.
Energy & Fuel, 22, 291-296.

GLAUDE P.A. (1999).
Construction automatique et validation de modèles cinétiques de combustion d'alcanes et d'éthers.
Thèse de l'Insitut National Polytechnique de Lorraine, Nancy.

GLAUDE P.A., BATTIN-LECLERC F., FOURNET R., WARTH V., CÔME
G.M., SCACCHI G. (2000).
Computer based generation of reaction mechanisms for gas-phase oxidation.
Combust. Flame, 122, 541-560.

GLAUDE P.A., CONRAUD V., FOURNET R., BATTIN-LECLERC F., CÔME G.M.,
SCACCHI G., DAGAUT P., CATHONNET M. (2002).
Modeling the oxidation of mixtures of primary reference fuels.
Energy & Fuels, 16, 1186-1195.

GLAUDE P.A., MARINOV N., KOSHIISHI Y., MATSUNAGA N., HORI M. (2005).
Kinetic modeling of the mutual oxidation of NO and larger alkanes at low temperature.
Energy & Fuels, 19, 1839-1849.

GUIBET J.C. (1997)
Carburants et Moteurs – Technologies – Energie – Environnement.
Editions Technip, ISBN 2-7108-0704-1.

HE Y., SANDERS W.A., LIN M.C. (1988).
Thermal decomposition of methyl nitrite: kinetic modeling of detailed product measurements by gas-liquid chromatography and Fourier transform infrared spectroscopy.
J. Phys. Chem., 92, 5474-5481.

HELDT T.H., CALLAHAN C.V., DRYER F.L. (1994).
The Effect of NO Addition on Methanol Oxidation at 12.5 atm, 700-820K.
Chem. Phys. Processes Combust., 270-273.

HELD T.J., MARCHESE A., DRYER F.L. (1997).
A Semi-Empirical Reaction Mechanism for n-Heptane Oxidation and Pyrolysis.
Combust. Sci. Tech., 123, 107-146.

HENRIOT S., BOUYSSOUNNOUSE D., BARITAUD T. (2003).
Port Fuel Injection and Combustion Simulation of a Racing Engine.
SAE Paper 2003-01-1845.

HERNANDEZ J.L., HERDING G., CARSTENSEN A. (2002)
A Study of the Thermochemical Conditions in the Exhaust Manifold Using Secondary Air in a 2.0-L Engine.
SAE, 2002-01-1676.

HERZLER J., FIKRI M., HITZBECK K., STARKE R., SCHULZ C., ROTH P., KALGHATGI G.T. (2007).
Shock Tube study of the autoignition of n-heptane/toluene/air mixtures at intermediate temperatures and high pressures.
Combust. Flame, 149, 25-31.

HEYBERGER B., BELMEKKI N., CONRAUD V., GLAUDE P.A., FOURNET R., BATTIN-LECLERC F. (2002).
Oxidation of small alkenes at high temperature.
Int. J. Chem. Kinet., 34, 666-677.

HEYWOOD J.B. (1988).
Internal Combustion Engine Fundamentals chap. 9.
McGraw-Hill, Inc., New York.

HORI M., MATSUNAGA N., MARINOV N., PITZ W.J., WESTBROOK C.K. (1998).
An Experimental and Kinetic Calculation of the Promotion Effect of Hydrocarbons on the NO-NO$_2$ Conversion in a Flow Reactor.
Proc. Combust. Inst., 27, 389-396.

HORI M., KOSHIISHI Y., MATSUNAGA N., GLAUDE P.A., MARINOV N. (2002).
Temperature dependence of NO to NO$_2$ conversion by n-butane and n-pentane oxidation.
Proc. Combust. Inst., 29, 2219-2226.

HUANG Y., SUNG C.J., ENG, J.A. (2004).
Laminar flame speeds of primary reference fuels and reformer gas mixtures.
Combust. Flame, 139, 239-251.

HUGHES K.J., TURÁNYI T., CLAGUE A.R., PILLING M.J. (2001).
Development and testing of a comprehensive chemical mechanism for the oxidation of methane.
Int. J. Chem. Kinet., 33, 513.

INGEMARSSON A.T., PEDERSEN J.R., OLSSON J.O. (1999).
Oxidation of n-heptane in a premixed flame.
J. Phys. Chem. A, 103, 8222-8230.

INGHAM T., WALKER R.W., WOLLFORD R.E., (1994).
Kinetic Parameters for the Initiation Reaction RH+ O$_2$-> R+ HO$_2$.
Proc. Combust. Inst., 25, 767-774.

JAY S., PERA C., COLIN O. (2007).
Livrables de la phase 2 de l'étude GSM E2.3 2006 – Tabulation de polluants par la méthode FPI et de la phase 3 de l'étude GSM DC1 2006 – FPI Moteur.
IFP Techniques d'Applications Energétiques, Centre de Résultats: Moteurs-Energie.

KEE R.J., RUPLEY F.M., MILLER J.A. (1993).
Sandia Report SAND89-8009B.
Sandia Laboratories Report.

KLEEMANN A.P., MENEGAZZI P., HENRIOT S., MARCHAL A. (2003).
Numerical Study on Knock and SI Engine by Thermally Coupling Combustion Chamber and Cooling Circuit Simulations.
AE Paper 2003-01-0563.

KLEEMANN A.P. (2006).
Etude GSM E4.2 2006 : Réduction des émissions lors de la mise en action – Potentiel de l'IAE.
IFP Techniques d'Applications Energétiques, Centre de Résultats: Moteurs-Energie.

KLOTZ S.D., BREZINSKY K., GLASSMAN I. (1998).
Modeling the Combustion of Toluene–Butane Blends.
Proc. Combust. Inst., 27, 337–344.

KOCHS W., KLODA, M., VENNE G. (2001).
Innovative Secondary Air Injection Systems.
SAE, 2001-01_0658.

KOROLL G.W., MULPURU S.R. (1986).
The Effect of Dilution with Steam on the Burning Velocity and Structure of Premixed Hydrogen Flames.
Twenty-First Symposium (International) on Combustion, 1811-1819.

KONG S.C., MARRIOTT C., REITZ R.D. (2001).
Modeling and Experiments of HCCI Engine Combustion Using Detailed Chemical Kinetics With Multidimensional CFD.
SAE, 2001-01-1026.

KONNOV A., BARNES F.J., BROMLEY J.H., ZHU J.N., ZHANG D. (2005).
A numerical study of the influence of ammonia addition on the auto-ignition limits of methane/air mixtures.
Combust. Flame, 141, 191-199.

LAFOSSAS F.A., CASTAGNE M., DUMAS J.P., HENRIOT S. (2002).
Development and Validation of a Knock Model in Spark Ignition Engines Using a CFD Code.
SAE Paper 2002-01-2701.

LAI H., THOMAS S. (1995).
Numerical Study of Contaminant Effects on Combustion of Hydrogen, Ethane and Methane in Air.
AIAA paper 95-6097, 12.

LAMOUREUX N., DJEBAILI-CHAUMEIX N., PAILLARD C.E. (2002).
Laminar flame velocity determination for H2-air-steam mixtures using the spherical bomb method.
Journal de Physique IV, 12, 445.

LAVOIE G.A., HEYWOOD J.B., KECK J.C. (1970).
Experimental and Theoretical Investigation of Nitric Oxide Formation in Internal Combustion Engines.
Combst. Sci. Technol., 1, 313-326.

LE CONG T. (2007).
Etude expérimentale et modélisation de la cinétique de combustion de combustibles gazeux :Méthane, gaz naturel et mélanges contenant de l'hydrogène, du monoxyde de carbone, du dioxyde de carbone et de l'eau.
Thèse de l'université d'Orléans.

LEPPARD W.R. (1992).
The autoignition chemistries of primary reference fuels, olefin/paraffin binary mixtures, and non-linear octane blending.
S.A.E. paper N°922325.

LI J., ZHAO Z.W., KAZAKOV A., DRYER F.L. (2004).
An updated comprehensive kinetic model of hydrogen combustion.
Int. J. Chem. Kinet., 36, 566.

LI J., ZHAO Z.W., KAZAKOV A., CHAOS M., DRYER F.L., SCIRE J.J. (2007).
A comprehensive kinetic mechanism for CO, CH2O and CH3OH combustion
Int. J. Chem. Kinet., 39, 109.

LIGHTFOOT P.D., COX R.A., CROWLEY J.N., DESTRIAU M. (1992).
Organic Peroxy Radicals: Kinetics, Spectroscopy and Troposheric Chemistry.
Atmos. Environ. A, 26, 1805–1961.

LINDSTEDT R.P., MAURICE L.Q. (1995).
Detailed Kinetic Modelling of n-heptane Combustion.
Combust. Sci. Technol., 107, 317–353.

LINDSTEDT R.P., MAURICE L.Q. (1996).
Detailed Kinetic Modelling of Toluene Combustion.
Combust. Sci. Technol., 120, 119–167.

LIU D.D.S.,MAC FARLANE R. (1983).
Laminar Burning Velocities of Hydrogen-Air and Hydrogen-Air-Steam Flames.
Combust. Flame, 49, 59-71.

LIU F., GUO H., SMALLWOOD G.J., GÜLDER Ö.L. (2001).
The Chemical Effects of Carbon Dioxide as an Additive in an Ethylene Diffusion Flame: Implications for Soot and NOx Formation.
Combust. Flame, 125, 778-787.

LÜ X.C., CHEN W., HUANG Z. (2005a).
A fundamental study on the control of the HCCI combustion and emissions by fuel design concept combined with controllable EGR. Part 1. The basic characteristics of HCCI combustion.
Fuel, 84, 1974-1083.

LÜ X.C., CHEN W., HUANG Z. (2005b).
A fundamental study on the control of the HCCI combustion and emissions by fuel design concept combined with controllable EGR. Part 2. Effect of operating conditions and EGR on HCCI combustion.
Fuel, 84, 1984-1092.

MACHRAFI H., CAVADIAS S., GUIBERT P. (2008).
An experimental and numerical investigation on the influence of external gas recirculation on the HCCI autoignition process in an engine: Thermal, diluting, and chemical effects.
Combust. Flame, 155, 476-489.

MARINOV N., PITZ W.J., WESTBROOK C.K., CASTALDI M.J., SENKAN S.M., (1996).
Modeling aromatic and polycyclic aromatic hydrocarbon formation in premixed methane and ethane flames.
Combust. Sci. Technol., 116-117, 211-287.

MARINOV N. (1998).
LLNL Report No. UCRL-JC-129372, Lawrence, Livermore National Laboratories, Berkeley, CA.

MAUVIOT G., ALBRECHT A., POINSOT T.J. (2006).
A new 0D approach for Diesel Combustion Modeling coupling probability density function with complex chemistry.
SAE 2006-01-3332.

MEBEL A.M., LIN M.C., MOROKUMA K. (1998).
Ab Inition MO and TST Calculations for the Rate Constant of the HNO + NO$_2$ HONO + NO Reaction.
Int. J. Chem. Kinet., 30, 729-736.

METGHALCHI M., KECK J.C. (1982).
Burning velocities of mixtures of air with methanol, isooctane and indolene at high pressure and temperature.
Combust. Flame 48, 191-210.

MINETTI R., CARLIER M., RIBAUCOUR M., THERSSEN E., SOCHET L.R. (1995).
A rapid compression machine investigation of oxidation and auto-ignition of n-Heptane: Measurements and modeling.
Combust. Flame, 102, 298-309.

MINETTI R., CARLIER M., RIBAUCOUR M., THERSSEN E., SOCHET L.R. (1996).
Comparison of Oxidation and Autoignition of the Two Primary Reference Fuels by Rapid Compression.
Proc. Combust. Inst. 26, 747-753.

MITANI T. (1995).
Ignition problems in scramjet testing.
Combust. Flame, 101, 347-359.

MITANI T., HIRAIWA T., SATO S., TOMIOKA S., KANDA T., TANI K. (1997).
Comparison of scramjet engine performance in Mach 6 Vitiated and Storage-Heated Air.
J. Prop. Power, 13, 635.

MITTAL G., SUNG C.J. (2007).
Autoignition of toluene and benzene at elevated pressures in a rapid compression machine.
Combust. Flame, 150, 355-368.

MIYOSHI A. (2005).
Development of an Auto-generation System for Detailed Kinetic Model of Combustion.
Jpn. Soc. Autom. Eng. 2005, 36, 35–40 (in Japanese).

MORÉAC G., DAGAUT P., ROESLER J.F., CATHONNET M. (2002).
Impact of Trace NO and other residual burnt gas components from piston engines on the oxidation of various hydrocarbons in a jet stirred reactor at atmospheric pressure.
Proc. 2nd Symp. (Med.) on Combust., The Combustion Institute, 2, 240-251.

MOREAC G. (2003).
Étude expérimentale et modélisation des interactions chimiques entre gaz résiduels et gaz frais dans l'allumage spontané homogène des moteurs à essence.
Thèse de l'Institut Français du Pétrole et du LCSR, Orléans.

MORÉAC G., DAGAUT P., ROESLER J., CATHONNET M. (2006).
Nitric oxide interactions with hydrocarbon oxidation in a jet-stirred reactor at 10 atm.
Combust. Flame, 145, 512-520.

MOTORLEXIKON (2009)
Abgasrückführung (AGR, EGR) ~äußere Abgasrückführung.
www.motorlexikon.de.

MUELLER M.A., YETTER R.A., DRYER F.L. (1999).
Flow reactor studies and kinetic modeling of the H2/O2/NOX and CO/H2O/O2/NOX reactions.
Int. J. Chem. Kinet. 31, 705.

MULLER C., MICHEL V., SCACCHI G., CÔME G.M. (1995).
THERGAS : a computer program for the evaluation of thermochemical data of molecules and free radicals in the gas phase.
J. Chim. Phys, 92, 1154-1178.

NAIK C.V., PITZ W.J., SJBERG M., DEC J.E., ORME J., CURRAN H.J., ET AL. (2005A)
Detailed chemical kinetic modelling of a surrogate fuels for gasolineand application to an HCCI engine.
In: Proceedings of the fourth joint meeting of the US sections of the Combustion Institute, Philadelphia.

NAIK C.V., PITZ W.J., SJBERG M., DEC J.E., ORME J., CURRAN H.J., SIMMIE J.M., WESTBROOK C.K. (2005B).
Detailed Chemical Kinetic Modeling of Surrogate Fuels for Gasoline and Application to an HCCI Engine.
SAE, 2005-01-3741.

NEHSE M., WARNATZ J. (1996).
Kinetic Modeling of the Oxidation of large aliphatic Hydrocarbons.
Proc. Combust. Inst. 26, 773-780.

NIST
National Institute for Standards and Technology.
www.nist.gov

OGURA T., SAKAI Y., MIYOSHI A., KOSHI M., DAGAUT P. (2007).
Modeling of the Oxidation of Primary Reference Fuel in the Presence of Oxygenated Octane Improvers: Ethyl Tert-Butyl Ether and Ethanol.
Energy & Fuels, 21, 3233-3239.

ONORATI A., FERRARI G., D'ERRICO G. (2003).
Secondary Air Injection in the Exhaust After-Treatment System of S.I. Engines: 1D Fluid Dynamic Modeling and Experimental Investigation.
SAE, 2003-01-0366.

PARK Y.M., CHOI I.V., DYAKOV M.C., LIN (2002).
An Experimental and Computational Study of the Thermal Oxidation of C6H5NO by NO2.
J. Phys. Chem. A 106, 2903-2907.

PERA C., COLIN O., JAY S. (2009).
Development of a FPI detailed chemistry tabulation methodology for internal combustion engines.
Oil & Gas Science and Technology – Rec. IFP, 57-1 &-16.

PILLING M.J. (1997).
Low-temperature Combustion and Autoignition.
Comprehensive Chemical Kinetics 35.

PELLETT G.L., NORTHAM G.B., WILSON L.G. (1992).
Strain-Induced Extinction of Hydrogen-Air Counterflow Diffusion Flames: Effects of Steam, CO$_2$, N$_2$ and O$_2$ Additives to Air.
AIAA Paper 92-0877, 15.

PIPEREL A., MONTAGNE X., DAGAUT P. (2007).
HCCI Engine Combustion Control Using EGR: Gas Composition Evolution and Consequences on Combustion Processes.
SAE, 2007-24-0087.

PIRES DA CRUZ A., PERA C., ANDERLOHR J., BOUNACEUR R., BATTIN-LECLERC F. (2007).
A complex chemical kinetic mechanism for the oxidation of gasoline surrogate fuels: n-heptane, iso-octane and toluene – Mechanism development and validation.
European Meeting of the Combustion Institute, Chania (Grece).

PITZ W.J., SEISER R., BOZZELLI J.W., DA COSTA I., FOURNET R., BILLAUD F., BATTIN-LECLERC F., SESHADRI K., WESTBROOK C.K. (2001).
Chemical characterisation of combustion of Toluene.
In: Proceedings of the second joint meeting of the US sections of the Combustion Institute, Oakland.

POINSOT T., VEYNANTE, D. (2005).
Theoretical and numerical combustion, 2nd ed. p.cm..
Edwards, ISBN 1-930217-10-2.

PRABHU S.K., BHAT R.K., MILLER D.L., CERNANSKY P. (1996).
*1-pentene Oxidation and its Interaction with Nitric oxide in the Low and Negative
Temperature Coefficient Regions.*
Combust. Flame, 104, 377-390.

RANZI E., SOGARO A., GAFFURI P., PENNATI G., FARAVELLI T. (1994).
A wide-range modeling study of methane oxidation.
Combust. Sci. and Technol. 96, 279.

RANZI E., GAFFURI P., FARAVELLI T., DAGAUT P. (1995).
A wide-range modeling study of n-heptane oxidation.
Combust. Flame 103, 91-106.

RANZI E., FARAVELLI T., GAFFURI P., SOGARO S., D'ANNA A., CIAJOLO A.
(1997).
A wide-range modeling study of iso-octane oxidation.
Combust. Flame, 108, 24-42.

ROUBAUD A., MINETTI R., SOCHET L.R. (2000).
*Oxidation and combustion of low alkylbenzenes at high pressure: comparative reactivity
and auto-ignition.*
Combust. Flame 123, 535–541.

RIBERT G., GICQUEL O., DARABIHA N., VEYNANTE D. (2006).
*Tabulation of complex chemistry based on sefl-similar behaviour of laminar premixed
flames.*
Combust. Flame 146, 640-664.

RISBERG.P., JOHANSSON D., ANDRAE J., KALGHATGI G., BJÖRNBOM P.,
ÄNGSTÖM H.E. (2006).
The Influence of NO on the Combustion Phasing in an HCCI Engine.
SAE, 2006-01-0416.

SAKAI Y., MIYOSHI A., MITSUO K., PITZ W. (2009)
*A kinetic modeling study on the oxidation of primary reference fuel-toluene mixtures
including cross reactions between aromatics and aliphatics.*
Proceedings of the Combustion Institute, 32, 411-418.

SASO Y. (2002).
Roles of Inhibitors in Global Gas-Phase Combustion kinetics.
Proceedings of the Combustion Institute, 29, 337-344.

SIVARAMAKRISHNAN R., TRANTER R.S., BREZINSKY K. (2004).
High-pressure, high-temperature oxidation of toluene.
Combust. Flame, 139, 340-350.

SJÖBERG, M., DEC J.E., HWANG, W. (2007)
Thermodynamic and Chemical Effects of EGR and Its Constituents on HCCI Autoingnition
SAE, 2007-01-0207.

SMITH G.P., GOLDEN D.M., FRENKLACH M., MORIARTY N. ET AL. (1999)
GRI-MECH 3.0.
http://www.me.berkeley.edu/gri_mech/

SOYHAN H.S., MAUSS F., SORUSBAY C. (2002).
Chemical Kinetic Modeling of Combustion in Internal Combustion Engines using reduced Chemistry.
Combust. Sci. Technol. 174, 73-91.

SRINIVASAN N.K., SU M.C., SUTHERLAND J.W., MICHAEL J.V. (2005).
Reflected shock tube studies of high-temperature rate constants for
$OH + CH_4 \text{--> } CH_3 + H_2O \text{ and } CH_3 + NO_2 \text{--> } CH_3O + NO$
J. Phys. Chem. A 109, 1857-1863.

SUBRAMANIAN G., BOUNACEUR R., PIRES DA CRUZ A., VERVISCH L. (2007).
Chemical impact of CO and H_2 addition on the auto-ignition delay of homogeneous n-heptane/air mixtures".
Combust. Sci. Technol. 197, 1937-1962.

TOUCHARD S. (2005).
Construction et validation de modèles cinétiques détaillés pour la combustion de mélanges modèles des essences.
Thèse de l'Institut National Polytechnique de Lorraine, Nancy.

TANAKA S., AYALA F., KECK J.C., HEYWOOD J.B. (2003a).
Two-stage ignition in HCCI combustion and HCCI control by fuels and additives.
Combust. Flame 132, 219-239.

TANAKA S., AYALA F., KECK J.C. (2003b).
A reduced chemical kinetic model for HCCI combustion of primary reference fuels in a rapid compression machine.
Combust. Flame 133, 467-481.

TROE J. (1974)
Fall-off curves of unimolecular reaction.
Ber. Buns. Phys. Chem. 78, 478.

TSANG W., HAMPSON R.F. (1986).
Chemical Kinetic Data Base for Combustion Chemistry. Part I. Methane and Related Compounds.
J. Phys. Chem. Ref. Data, 15, 1087.

TSANG W. (1987).
Chemical kinetic data base for combustion chemistry. Part 2. Methanol.
J. Phys. Chem. Ref. Data 16, 471-508.

TSANG W., HERRON J.T. (1991).
Chemical Kinetic Data Base for Propellant Combustion I. Reactions Involving NO, NO$_2$, HNO, HNO$_2$, HCN and N$_2$O.
J. Phys. Chem. Ref. Data 20, 609-663.

TSENG C.M., CHOI Y.M., HUANG C.L., NI C.K., LEE Y.T., LIN M.C. (2004).
Photodissociation of Nitrosobenzene and Decomposition of Phenyl Radical.
J. Phys. Chem. A 108 7928-7935.

VANHOVE G., PETIT G., MINETTI R. (2006).
Experimental study of the kinetic interactions in the low-temperature autoignition of hydrocarbon binary mixtures and a surrogate fuel.
Combust. Flame 145, 521-532.

WARNATZ J. (1984).
Chemistry of high temperature combustion of alkanes up to octane.
Twentieth Symposium (International) on Combustion, 845-856.

WALKER R.W., MORLEY C. (1997).
Basic chemistry of combustion: In Comprehensive Chemical Kinetics: low-temperature combustion and autoignition..
Pilling MJ Ed., 35, Elsevier, Amsterdam, 1997.

WALLINGTON T.J., DAGAUT P., KURYLO M.J. (1992).
Ultraviolet absorption cross sections and reaction kinetics and mechanisms for peroxy radicals in the gas phase.
Chem. Rev., 92, 667.

WANG B.L., OLIVIER H., GRÖNIG H. (2003).
Ignition of shock-heated H2-air-steam mixtures.
Combust. Flame, 133, 93.

WARTH V., STEF N., GLAUDE P.A., BATTIN-LECLERC F., SCACCHI G., COME G.M. (1998).
Computer-Aided Derivation of Gas-Phase Oxidation Mechanisms: Application to the Modeling of the Oxidation of n-Butane.
Combust. Flame, 114, 81-102.

WESTBROOK C.K., DRYER F.L. (1984).
Chemical kinetic modelling of hydrocarbon combustion.
Prog. Energy Combust. Sci., 10, 1.

WESTBROOK C.K., WARNATZ J., PITZ W.J. (1988).
A detailed Chemical Kinetic Reaction Mechanism for the Oxidation of Iso-octane and n-Heptane over an Extended Temperature Range and its Application to Analysis of Engine Knock.
Proc. Combust. Inst., 22, 893-901.

WESTBROOK C.K. (2000).
Chemical Kinetics of Hydrocarbon Ignition in Practical Combustion Systems.
Proceedings of the Combustion Institute, 28, 1563-1577.

XU S., LIN M.C. (2003).
Kinetics and mechanism for the CH2O + NO2 reaction: A computational study
Int. J. Chem. Kinet. 35, 184-190.

XU S., LIN M.C. (2005).
J. Phys. Chem. B 109, 8367-8373.

YAHAYAOUI M., DJEBAILI-CHAUMEIX N., DAGAUT P., PAILLARD C.E., GAIL S. (2007).
Experimental and modelling study of gasoline surrogate mixtures oxidation in jet stirred reactor and shock tube.
Proc. Combust. Inst., 31, 385-391.

YETTER R.A., DRYER F.L., RABITZ H. (1991a).
A comprehensive reaction mechanism for carbon-monoxide hydrogen oxygen kinetics.
Combust. Sci. and Technol., 79, 97.

YETTER R.A., DRYER F.L., RABITZ H. (1991b).
Flow reactor studies of carbon-monoxide hydrogen oxygen kinetics.
Combust. Sci. and Technol., 79, 129.

ZELDOVICH Y.B. (1946).
The Oxidation of Nitrogen in Combustion and Explosions.
Acta Physiochim. U.R.S.S. *21:577-628.*

ZSÉLY.I., ZÁDOR J., TURÁNYI T. (2005).
Uncertainty analysis of updated hydrogen and carbon monoxide oxidation mechanisms.
Proc. Combust. Inst., 30, 1273.

ZHENG J., YANG W., MILLER D.L., CERNANSKY N.P. (2001).
Prediction of Pre-ignition Reactivity and Ignition Delay for HCCI Using a Reduced Chemical Kinetic Model.
SAE, 2001-01-1025.

APPENDIX

Appendix A

Modelling PRF/toluene/NO_x reaction kinetics

A.1 Reaction mechanism of NO_x species

The reactions between hydrocarbons and NOx-specie are shown here:

```
!***************************************************************************
!*                                                                        *
!*          GRI-MECH3.0 C0-C2-NOX                                         *
!*                                                                        *
!***************************************************************************

! NITROGEN CHEMISTRY FROM HORI ET AL. (27TH SYMPOSIUM) AND FROM MARINOV
N+NO=N2+B1O                       2.700E+13   .000    355.00 ! GRI-MECH3.0
N+O2=NO+B1O                       9.000E+09  1.000   6500.00 ! GRI-MECH3.0
N+R2OH=NO+R1H                     3.360E+13   .000    385.00 ! GRI-MECH3.0
N2O+B1O=N2+O2                     1.40E+12    0.00   10810.0 ! GRI-MECH3.0
N2O+B1O=2NO                       2.90E+13    0.00   23150.0 ! GRI-MECH3.0
N2O+R1H=N2+R2OH                   3.870E+14   .000   18880.00 ! GRI-MECH3.0
N2O+R2OH=N2+R3OOH                 2.000E+12   .000   21060.00 ! GRI-MECH3.0
N2O(+M)=N2+B1O(+M)                7.910E+10   .000   56020.00 ! GRI-MECH3.0
    LOW  / 6.370E+14   .000 56640.00/
H2/2.00/ H2O/6.00/ CH4/2.00/ B2CO/1.50/ CO2/2.00/ C2H6/3.00/ AR/ .625/
NO+R3OOH=NO2+R2OH                 2.110E+12   .000   -480.00 ! GRI-MECH3.0
NO+B1O+M=NO2+M                    1.060E+20  -1.410      .00 ! GRI-MECH3.0
    H2/2.00/ H2O/6.00/ CH4/2.00/ B2CO/1.50/ CO2/2.00/ C2H6/3.00/ AR/ .70/
NO2+B1O=NO+O2                     3.900E+12   .000   -240.00 ! GRI-MECH3.0
NO2+R1H=NO+R2OH                   1.320E+14   .000    360.00 ! GRI-MECH3.0
NH+B1O=NO+R1H                     4.000E+13   .000      .00 ! GRI-MECH3.0
NH+R1H=N+H2                       3.200E+13   .000    330.00 ! GRI-MECH3.0
NH+R2OH=HNO+R1H                   2.000E+13   .000      .00 ! GRI-MECH3.0
NH+R2OH=N+H2O                     2.000E+09  1.200      .00 ! GRI-MECH3.0
NH+O2=HNO+B1O                     4.610E+05  2.000   6500.00 ! GRI-MECH3.0
NH+O2=NO+R2OH                     1.280E+06  1.500    100.00 ! GRI-MECH3.0
NH+N=N2+R1H                       1.500E+13   .000      .00 ! GRI-MECH3.0
NH+H2O=HNO+H2                     2.000E+13   .000  13850.00 ! GRI-MECH3.0
NH+NO=N2+R2OH                     2.160E+13  -.230      .00 ! GRI-MECH3.0
NH+NO=N2O+R1H                     3.650E+14  -.450      .00 ! GRI-MECH3.0
NH2+B1O=R2OH+NH                   3.000E+12   .000      .00 ! GRI-MECH3.0
NH2+B1O=R1H+HNO                   3.900E+13   .000      .00 ! GRI-MECH3.0
NH2+R1H=NH+H2                     4.000E+13   .000   3650.00 ! GRI-MECH3.0
NH2+R2OH=NH+H2O                   9.000E+07  1.500   -460.00 ! GRI-MECH3.0
NNH=N2+R1H                        3.300E+08   .000      .00 ! GRI-MECH3.0
    ! DUP
NNH+M=N2+R1H+M                    1.30E+14   -.110   4980.00 ! GRI-MECH3.0
    ! DUP
H2/2.00/ H2O/6.00/ CH4/2.00/ B2CO/1.50/ CO2/2.00/ C2H6/3.00/ AR/0.70/
NNH+O2=R3OOH+N2                   5.000E+12   .000      .00 ! GRI-MECH3.0
NNH+B1O=R2OH+N2                   2.500E+13   .000      .00 ! GRI-MECH3.0
NNH+B1O=NH+NO                     7.000E+13   .000      .00 ! GRI-MECH3.0
NNH+R1H=H2+N2                     5.000E+13   .000      .00 ! GRI-MECH3.0
NNH+R2OH=H2O+N2                   2.000E+13   .000      .00 ! GRI-MECH3.0
NNH+R4CH3=CH4+N2                  2.500E+13   .000      .00 ! GRI-MECH3.0
R1H+NO+M=HNO+M                    8.95E+19   -1.320    740.00 !GRI-MECH2.11
H2/2.00/ H2O/6.00/ CH4/2.00/ B2CO/1.50/ CO2/2.00/ C2H6/3.00/ AR/0.70/
HNO+B1O=NO+R2OH                   2.500E+13   .000      .00 ! GRI-MECH3.0
HNO+R1H=H2+NO                     9.000E+11   .720    660.00 !GRI-MECH3.0
HNO+R2OH=NO+H2O                   1.300E+07  1.900   -950.00 !GRI-MECH3.0
CN+B1O<=>B2CO+N                   7.700E+13   .000      .00 !GRI-MECH3.0
CN+R2OH<=>NCO+R1H                 4.000E+13   .000      .00 !GRI-MECH3.0
CN+H2O<=>HCN+R2OH                 8.000E+12   .000   7460.00  !GRI-MECH3.0
CN+O2<=>NCO+B1O                   6.140E+12   .000   -440.00  !GRI-MECH3.0
CN+H2<=>HCN+R1H                   2.950E+05  2.450   2240.00  !GRI-MECH3.0
NCO+B1O<=>NO+B2CO                 2.350E+13   .000      .00 !GRI-MECH3.0
NCO+R1H<=>NH+B2CO                 5.400E+13   .000      .00 !GRI-MECH3.0
NCO+R2OH<=>NO+R1H+B2CO            0.250E+13   .000      .00 !GRI-MECH3.0
NCO+N<=>N2+B2CO                   2.000E+13   .000      .00  !GRI-MECH3.0
NCO+O2<=>NO+CO2                   2.000E+12   .000  20000.00   !GRI-MECH3.0
NCO+M<=>N+B2CO+M                  3.100E+14   .000  54050.00   !GRI-MECH3.0
H2/2.00/ H2O/6.00/ CH4/2.00/ B2CO/1.50/ CO2/2.00/ C2H6/3.00/ AR/ .70/
NCO+NO<=>N2O+B2CO                 1.900E+17  -1.520    740.00   !GRI-MECH3.0
NCO+NO<=>N2+CO2                   3.800E+18  -2.000    800.00   !GRI-MECH3.0
```

```
HCN+M<=>R1H+CN+M                1.040E+30  -3.300 126600.00   !GRI-MECH3.0
H2/2.00/ H2O/6.00/ CH4/2.00/ B2CO/1.50/ CO2/2.00/ C2H6/3.00/ AR/ .70/      !GRI-MECH3.0
HCN+B1O<=>NCO+R1H               2.030E+04   2.640  4980.00 !GRI-MECH3.0
HCN+B1O<=>NH+B2CO               5.070E+03   2.640  4980.00 !GRI-MECH3.0
HCN+B1O<=>CN+R2OH               3.910E+09   1.580 26600.00 !GRI-MECH3.0
HCN+R2OH<=>HOCN+R1H             1.100E+06   2.030 13370.00 !GRI-MECH3.0
HCN+R2OH<=>HNCO+R1H             4.400E+03   2.260  6400.00 !GRI-MECH3.0
HCN+R2OH<=>NH2+B2CO             1.600E+02   2.560  9000.00 !GRI-MECH3.0
R1H+HCN(+M)<=>H2CN(+M)          3.300E+13   .000     .00  !GRI-MECH3.0
   LOW / 1.400E+26  -3.400   1900.00/
H2/2.00/ H2O/6.00/ CH4/2.00/ B2CO/1.50/ CO2/2.00/ C2H6/3.00/ AR/ .70/
H2CN+N<=>N2+B5CH2               6.000E+13   .000   400.00 !GRI-MECH3.0
B3C+N2<=>CN+N                   6.300E+13   .000 46020.00 !GRI-MECH3.0
B4CH+N2<=>HCN+N                 3.120E+09   0.880 20130.00 !GRI-MECH3.0
B4CH+N2(+M)<=>HCNN(+M)          3.100E+12   .150     .00 !GRI-MECH3.0
   LOW / 1.300E+25  -3.160    740.00/
   TROE/ .6670 235.00 2117.00 4536.00 /
H2/2.00/ H2O/6.00/ CH4/2.00/ B2CO/1.50/ CO2/2.00/ C2H6/3.00/ AR/ 1.0/
B5CH2+N2<=>HCN+NH               1.000E+13   .000 74000.00 !GRI-MECH3.0
B6CH2+N2<=>NH+HCN               1.000E+11   .000 65000.00   !GRI-MECH3.0
B3C+NO<=>CN+B1O                 1.900E+13   .000     .00 !GRI-MECH3.0
B3C+NO<=>B2CO+N                 2.900E+13   .000     .00 !GRI-MECH3.0
B4CH+NO<=>HCN+B1O               4.100E+13   .000     .00 !GRI-MECH3.0
B4CH+NO<=>R1H+NCO               1.620E+13   .000     .00 !GRI-MECH3.0
B4CH+NO<=>N+R5CHO               2.460E+13   .000     .00 !GRI-MECH3.0
B5CH2+NO<=>R1H+HNCO             3.100E+17  -1.380  1270.00 !GRI-MECH3.0
B5CH2+NO<=>R2OH+HCN             2.900E+14   -.690   760.00 !GRI-MECH3.0
B5CH2+NO<=>R1H+HCNO             3.800E+13   -.360   580.00 !GRI-MECH3.0
B6CH2+NO<=>R1H+HNCO             3.100E+17  -1.380  1270.00 !GRI-MECH3.0
B6CH2+NO<=>R2OH+HCN             2.900E+14   -.690   760.00 !GRI-MECH3.0
B6CH2+NO<=>R1H+HCNO             3.800E+13   -.360   580.00 !GRI-MECH3.0
R4CH3+NO<=>HCN+H2O              9.600E+13   .000 28800.00 !GRI-MECH3.0
R4CH3+NO<=>H2CN+R2OH            1.000E+12   .000 21750.00 !GRI-MECH3.0
HCNN+B1O<=>B2CO+R1H+N2          2.200E+13   .000     .00 !GRI-MECH3.0
HCNN+B1O<=>HCN+NO               2.000E+12   .000    .00  !GRI-MECH3.0
HCNN+O2<=>B1O+R5CHO+N2          1.200E+13   .000    .00  !GRI-MECH3.0
HCNN+R2OH<=>R1H+R5CHO+N2        1.200E+13   .000    .00  !GRI-MECH3.0
HCNN+R1H<=>B5CH2+N2             1.000E+14   .000    .00  !GRI-MECH3.0
HNCO+B1O<=>NH+CO2              9.800E+07   1.410  8500.00   !GRI-MECH3.0
HNCO+B1O<=>HNO+B2CO            1.500E+08   1.570 44000.00 !GRI-MECH3.0
HNCO+B1O<=>NCO+R2OH            2.200E+06   2.110 11400.00 !GRI-MECH3.0
HNCO+R1H<=>NH2+B2CO            2.250E+07   1.700  3800.00 !GRI-MECH3.0
HNCO+R1H<=>H2+NCO             1.050E+05   2.500 13300.00 !GRI-MECH3.0
HNCO+R2OH<=>NCO+H2O           3.300E+07   1.500  3600.00 !GRI-MECH3.0
HNCO+R2OH<=>NH2+CO2           3.300E+06   1.500  3600.00 !GRI-MECH3.0
HNCO+M<=>NH+B2CO+M            1.180E+16   .000 84720.00 !GRI-MECH3.0
H2/2.00/ H2O/6.00/ CH4/2.00/ B2CO/1.50/ CO2/2.00/ C2H6/3.00/ AR/ .70/
HCNO+R1H<=>R1H+HNCO           2.100E+15   -.690  2850.00 !GRI-MECH3.0
HCNO+R1H<=>R2OH+HCN           2.700E+11   .180  2120.00 !GRI-MECH3.0
HCNO+R1H<=>NH2+B2CO           1.700E+14   -.750  2890.00 !GRI-MECH3.0
HOCN+R1H<=>R1H+HNCO           2.000E+07   2.000  2000.00 !GRI-MECH3.0
R12CHCOV+NO<=>HCNO+B2CO       0.900E+13   .000    .00 !GRI-MECH3.0
R4CH3+N<=>H2CN+R1H            6.100E+14   -.310   290.00 !GRI-MECH3.0
R4CH3+N<=>HCN+H2              3.700E+12   .150   -90.00 !GRI-MECH3.0
NH3+R1H<=>NH2+H2              5.400E+05   2.400  9915.00   !GRI-MECH3.0
NH3+R2OH<=>NH2+H2O            5.000E+07   1.600   955.00   !GRI-MECH3.0
NH3+B1O<=>NH2+R2OH            9.400E+06   1.940  6460.00  !GRI-MECH3.0
NH+CO2<=>HNO+B2CO             1.000E+13   .000 14350.00 ! GRI-MECH3.0
CN+NO2<=>NCO+NO              6.160E+15  -0.752   345.00 ! GRI-MECH3.0
NCO+NO2<=>N2O+CO2            3.250E+12   .000  -705.00 ! GRI-MECH3.0
N+CO2<=>NO+B2CO              3.000E+12   .000 11300.00 ! GRI-MECH3.0
```

!REACTIONS MODIFIED BY ANDELROHR COMPARED TO THE GRI-MECH3.0
```
HNO+O2<=>R3OOH+NO             8.0E10   .000  9520.00  ! *3.6 BRYUKOV93
```

```
!********************************************************************
!*                                                                *
!*      REACTIONS ADDED BY GLAUDE                                  *
!*                                                                *
!********************************************************************
CH3NO2+R2OH=HCHO+NO+H2O       3.00E+6    2.0   2000.0 !THIS STUDY
CH3NO2+B1O=HCHO+NO+R2OH       1.51E+13   0.0   5354.0 !SALTER 1977
CH3NO2+R1H=HCHO+NO+H2         4.67E+12   0.0   3732.0 !KO 1991
```

```
CH3NO2+R4CH3=HCHO+NO+CH4        7.08E+11   0.0   11140.0 !BALLOD 1980
HNO2+R1H=NO2+H2                 2.40E+8    1.5    5087.0 !DEAN 1997
HNO2+B1O=NO2+R2OH              1.70E+8    1.5    3020.0 !DEAN 1997
HNO2+R2OH=NO2+H2O              1.20E+6    2.0    -596.0 !DEAN 1997
HNO2+R4CH3=NO2+CH4             8.10E+5    1.87   4838.0 !DEAN 1997
HNO2=HONO                      1.30E+29  -5.47  52814.0 !DEAN 1997
HONO+R4CH3=NO2+CH4             8.10E+5    1.87   5504.0 !DEAN 1997
HONO+R2OH=NO2+H2O             1.26E+10    1.0    135.0 !DEAN 1997
HONO+B1O=NO2+R2OH            1.21E+13    0.0    5962.0 !DEAN 1997
HONO+R1H=NO2+H2             1.21E+13    0.0    7353.0 !DEAN 1997
HONO+HONO=NO+NO2+H2O         1.02E+11    0.0    8540.0 !ENgLAND 1975
R2OH+NO2(+M)=HONO2(+M)        2.40E+13    0.0      0.0 !TSANg 1991
   LOw / 6.42E+32 -5.49 2351./
   TROE/0.525 1.0E-15 1.0E-15 1.0E+15/H2O /5.0/
HONO2+R2OH=NO3+H2O            1.03E+10    0.0   -1240.0 !LAMB 1984
NO3=NO+O2                     2.50E+6     0.0   12122.0 !JOHNSTON 1986
NO2+B1O+M=NO3+M               2.94E+21   -2.0      0.0 !ATkINSON 1989
NO2+NO2=NO3+NO                9.60E+9    0.73  20923.0 !TSANg 1991
   DUPLICATE
NO2+NO2=NO3+NO                1.60E+12    0.0   26123.0 !TSANg 1991
   DUPLICATE
R7CH3O+NO(+M)=CH3ONO(+M)     1.21E+13    0.0    -322.0 !ATkINSON 1992
   LOw / 2.70E+27 -3.50 0.0/
R8CH3OO+NO=R7CH3O+NO2                2.53E+12  0.0   -358.0 !ATkINSON 1992
R17C2H5OO+NO=R15C2H5O+NO2            2.53E+12  0.0   -358.0 !STUDY GLAUDE
R29C3H7OO+NO=R11C2H5+HCHO+NO2        2.11E+12  0.0   -358.0 !STUDY GLAUDE
R15C2H5O+NO=CH3CHO+HNO               4.00E+13 -0.6      0.0 !STUDY GLAUDE
R4CH3+NO(+M)=CH3NO(+M)              2.17E+11   0.6      0.0 !JODkOwSkI 1993
   LOw /2.06E+27 -3.50 0.0/
B2CO+NO2=CO2+NO                      9.04E+13  0.0   33782.0 !TSANg 1991
R7CH3O+NO2(+M)=CH3ONO2(+M)          1.20E+13  0.0      0.0 !ATkINSON 1992
   LOw / 1.40E+30 -4.50 0.0/
R11C2H5+NO2=R15C2H5O+NO         1.36E+13  0.0      0.0 !comme SRINIVASAN05 R4CH3+NO2=R7CH3O+NO
R13CH2CHO+NO2=HCHO+R5CHO+NO     1.00E+13  0.0      0.0 !BARNHARD 1991
RC3H5Y+NO2=>R10C2H3V+HCHO+NO    2.30E+13  0.0      0.0 !SLAgLE 1981
NO2+R3OOH=HONO+O2              3.65E+13  0.0    8000.0 !STUDY GLAUDE
R7CH3O+NO2=HCHO+HONO          4.00E+12  0.0    2285.0 !
R15C2H5O+NO2=CH3CHO+HONO      2.70E+12  0.0     105.0 !STUDY GLAUDE
```

```
!******************************************************************************
!*                                                                           *
!*   REACTIONS ADDED OR MODIFIED COMPARED TO GLAUDE                          *
!*                                                                           *
!******************************************************************************
```

```
NO+R2OH(+M)=HONO(+M)            1.1E+14   -0.3      0.0 !ATkINSON04
   LOw /2.35E+23 -2.4 0.0/       !ATKINSON04
   SRI /1.0 0.0 1.E-18 0.81 0.0/
!Fall off Parameter Fc =0.81
R4CH3+NO2=R7CH3O+NO            1.36E+13  0.0      0.0 !SRINIVASAN05
!CH3NO2(+M)=R4CH3+NO2(+M)      1.78E+16  0.0   58500.0 !GLAENZER 1972
!    LOw /1.26E+17 0.0 42000./
CH3NO2(+M)=R4CH3+NO2(+M)       1.8E+17   0.0   58500.0 !
   LOw /1.3E+18 0.0 42000./
R7CH3O+NO=HCHO+HNO             7.60E+13 -0.76     0.0 ! ATKINSON05
HCHO+NO=R5CHO+HNO             1.02e13 0.0  40670.0 ! TSANG1991
HCHO+NO2=R5CHO+HONO          8.35E-11 6.68 8300! XU05
HNO+NO2=HNO2+NO              3.0E11   0.0 1988 ! TSANG1991
HNO+NO=R2OH+N2O             8.5E12   0.0 29640 ! DIAU1995
HNO+R4CH3=CH4+NO           1.47E11  0.76  349 !CHOI2005
HNO+R7CH3O=CH3OH+NO        3.16E13  0.0  0.0 !HE1988
HONO+NO=NO2+HNO            4.4E+3   2.64  4038.0 !Mebel1998
```

```
!******************************************************************************
!*                                                                           *
!*         COUPLING OF HC(nC) > 3 WITH NOx                                    *
!*                                                                           *
!******************************************************************************
```

```
! REACTIONS : ROO + NO = NO2 + RO
R29C3H7OO + NO => R1C3H7O + NO2      4.7000e12 0.0 -358.0 ! STUDY ANDERLOHR
R166C3H7OO + NO => R2C3H7O + NO2     4.7000e12 0.0 -358.0 ! STUDY ANDERLOHR
R30C4H9OO + NO => R1C4H9O + NO2      4.7000e12 0.0 -358.0 ! STUDY ANDERLOHR
```

```
R72C4H9OO + NO =>  R2C4H9O + NO2          4.7000e12 0.0 -358.0 ! STUDY ANDERLOHR
R81C4H9OO + NO =>  R3C4H9O + NO2          4.7000e12 0.0 -358.0 ! STUDY ANDERLOHR
R168C4H9OO + NO =>  R4C4H9O + NO2         4.7000e12 0.0 -358.0 ! STUDY ANDERLOHR
R55C5H11OO + NO =>  R1C5H11O + NO2        4.7000e12 0.0 -358.0 ! STUDY ANDERLOHR
R77C5H11OO + NO =>  R2C5H11O + NO2        4.7000e12 0.0 -358.0 ! STUDY ANDERLOHR
R107C5H11OO + NO => R3C5H11O + NO2        4.7000e12 0.0 -358.0 ! STUDY ANDERLOHR
R174C5H11OO + NO => R4C5H11O + NO2        4.7000e12 0.0 -358.0 ! STUDY ANDERLOHR
R31C7H15OO + NO =>  R1C7H15O + NO2        4.7000e12 0.0 -358.0 ! STUDY ANDERLOHR
R32C7H15OO + NO =>  R2C7H15O + NO2        4.7000e12 0.0 -358.0 ! STUDY ANDERLOHR
R34C7H15OO + NO =>  R3C7H15O + NO2        4.7000e12 0.0 -358.0 ! STUDY ANDERLOHR
R35C7H15OO + NO =>  R4C7H15O + NO2        4.7000e12 0.0 -358.0 ! STUDY ANDERLOHR
R36C8H17OO + NO =>  R1C8H17O + NO2        4.7000e12 0.0 -358.0 ! STUDY ANDERLOHR
R38C8H17OO + NO =>  R2C8H17O + NO2        4.7000e12 0.0 -358.0 ! STUDY ANDERLOHR
R40C8H17OO + NO =>  R3C8H17O + NO2        4.7000e12 0.0 -358.0 ! STUDY ANDERLOHR
R42C8H17OO + NO =>  R4C8H17O + NO2        4.7000e12 0.0 -358.0 ! STUDY ANDERLOHR
```

! DECOMPOSITION REACTIONS : OOQOOH + NO = NO2 + OH + HCHO + olefin
! Decompostion de la facon générique suivante:
! RCnH(2n)OOOOH + NO => NO2 + R2OH + 2HCHO + C(n-2)H(2n-4)

```
R85C3H6OOOOH + NO   => NO2 + R2OH  + 2HCHO + B6CH2        4.7000e12 0.0 -358.0 ! STUDY ANDERLOHR
R86C3H6OOOOH + NO   => NO2 + R2OH  + 2HCHO + B6CH2        4.7000e12 0.0 -358.0 ! STUDY ANDERLOHR
R165C3H6OOOOH + NO  => NO2 + R2OH  + 2HCHO + B6CH2        4.7000e12 0.0 -358.0 ! STUDY ANDERLOHR
R170C3H6OOOOH + NO  => NO2 + R2OH  + 2HCHO + B6CH2        4.7000e12 0.0 -358.0 ! STUDY ANDERLOHR
R87C4H8OOOOH + NO   => NO2 + R2OH  + 2HCHO + C2H4Z        4.7000e12 0.0 -358.0 ! STUDY ANDERLOHR
R88C4H8OOOOH + NO   => NO2 + R2OH  + 2HCHO + C2H4Z        4.7000e12 0.0 -358.0 ! STUDY ANDERLOHR
R89C4H8OOOOH + NO   => NO2 + R2OH  + 2HCHO + C2H4Z        4.7000e12 0.0 -358.0 ! STUDY ANDERLOHR
R150C4H8OOOOH + NO  => NO2 + R2OH  + 2HCHO + C2H4Z        4.7000e12 0.0 -358.0 ! STUDY ANDERLOHR
R167C4H8OOOOH + NO  => NO2 + R2OH  + 2HCHO + C2H4Z        4.7000e12 0.0 -358.0 ! STUDY ANDERLOHR
R172C4H8OOOOH + NO  => NO2 + R2OH  + 2HCHO + C2H4Z        4.7000e12 0.0 -358.0 ! STUDY ANDERLOHR
R177C4H8OOOOH + NO  => NO2 + R2OH  + 2HCHO + C2H4Z        4.7000e12 0.0 -358.0 ! STUDY ANDERLOHR
R179C4H8OOOOH + NO  => NO2 + R2OH  + 2HCHO + C2H4Z        4.7000e12 0.0 -358.0 ! STUDY ANDERLOHR
R181C4H8OOOOH + NO  => NO2 + R2OH  + 2HCHO + C2H4Z        4.7000e12 0.0 -358.0 ! STUDY ANDERLOHR
R183C4H8OOOOH + NO  => NO2 + R2OH  + 2HCHO + C2H4Z        4.7000e12 0.0 -358.0 ! STUDY ANDERLOHR
R194C4H8OOOOH + NO  => NO2 + R2OH  + 2HCHO + C2H4Z        4.7000e12 0.0 -358.0 ! STUDY ANDERLOHR
R152C5H10OOOOH + NO => NO2 + R2OH  + 2HCHO + C3H6Y        4.7000e12 0.0 -358.0 ! STUDY ANDERLOHR
R158C5H10OOOOH + NO => NO2 + R2OH  + 2HCHO + C3H6Y        4.7000e12 0.0 -358.0 ! STUDY ANDERLOHR
R159C5H10OOOOH + NO => NO2 + R2OH  + 2HCHO + C3H6Y        4.7000e12 0.0 -358.0 ! STUDY ANDERLOHR
R160C5H10OOOOH + NO => NO2 + R2OH  + 2HCHO + C3H6Y        4.7000e12 0.0 -358.0 ! STUDY ANDERLOHR
R161C5H10OOOOH + NO => NO2 + R2OH  + 2HCHO + C3H6Y        4.7000e12 0.0 -358.0 ! STUDY ANDERLOHR
R169C5H10OOOOH + NO => NO2 + R2OH  + 2HCHO + C3H6Y        4.7000e12 0.0 -358.0 ! STUDY ANDERLOHR
R173C5H10OOOOH + NO => NO2 + R2OH  + 2HCHO + C3H6Y        4.7000e12 0.0 -358.0 ! STUDY ANDERLOHR
R178C5H10OOOOH + NO => NO2 + R2OH  + 2HCHO + C3H6Y        4.7000e12 0.0 -358.0 ! STUDY ANDERLOHR
R182C5H10OOOOH + NO => NO2 + R2OH  + 2HCHO + C3H6Y        4.7000e12 0.0 -358.0 ! STUDY ANDERLOHR
R184C5H10OOOOH + NO => NO2 + R2OH  + 2HCHO + C3H6Y        4.7000e12 0.0 -358.0 ! STUDY ANDERLOHR
R185C5H10OOOOH + NO => NO2 + R2OH  + 2HCHO + C3H6Y        4.7000e12 0.0 -358.0 ! STUDY ANDERLOHR
R187C5H10OOOOH + NO => NO2 + R2OH  + 2HCHO + C3H6Y        4.7000e12 0.0 -358.0 ! STUDY ANDERLOHR
R192C5H10OOOOH + NO => NO2 + R2OH  + 2HCHO + C3H6Y        4.7000e12 0.0 -358.0 ! STUDY ANDERLOHR
R193C5H10OOOOH + NO => NO2 + R2OH  + 2HCHO + C3H6Y        4.7000e12 0.0 -358.0 ! STUDY ANDERLOHR
R91C7H14OOOOH + NO  => NO2 + R2OH  + 2HCHO + C5H10Z       4.7000e12 0.0 -358.0 ! STUDY ANDERLOHR
R92C7H14OOOOH + NO  => NO2 + R2OH  + 2HCHO + C5H10Z       4.7000e12 0.0 -358.0 ! STUDY ANDERLOHR
R93C7H14OOOOH + NO  => NO2 + R2OH  + 2HCHO + C5H10Z       4.7000e12 0.0 -358.0 ! STUDY ANDERLOHR
R96C7H14OOOOH + NO  => NO2 + R2OH  + 2HCHO + C5H10Z       4.7000e12 0.0 -358.0 ! STUDY ANDERLOHR
R99C7H14OOOOH + NO  => NO2 + R2OH  + 2HCHO + C5H10Z       4.7000e12 0.0 -358.0 ! STUDY ANDERLOHR
R100C7H14OOOOH + NO => NO2 + R2OH  + 2HCHO + C5H10Z       4.7000e12 0.0 -358.0 ! STUDY ANDERLOHR
R102C7H14OOOOH + NO => NO2 + R2OH  + 2HCHO + C5H10Z       4.7000e12 0.0 -358.0 ! STUDY ANDERLOHR
R108C7H14OOOOH + NO => NO2 + R2OH  + 2HCHO + C5H10Z       4.7000e12 0.0 -358.0 ! STUDY ANDERLOHR
R110C7H14OOOOH + NO => NO2 + R2OH  + 2HCHO + C5H10Z       4.7000e12 0.0 -358.0 ! STUDY ANDERLOHR
R111C7H14OOOOH + NO => NO2 + R2OH  + 2HCHO + C5H10Z       4.7000e12 0.0 -358.0 ! STUDY ANDERLOHR
R112C7H14OOOOH + NO => NO2 + R2OH  + 2HCHO + C5H10Z       4.7000e12 0.0 -358.0 ! STUDY ANDERLOHR
R115C7H14OOOOH + NO => NO2 + R2OH  + 2HCHO + C5H10Z       4.7000e12 0.0 -358.0 ! STUDY ANDERLOHR
R117C7H14OOOOH + NO => NO2 + R2OH  + 2HCHO + C5H10Z       4.7000e12 0.0 -358.0 ! STUDY ANDERLOHR
R118C7H14OOOOH + NO => NO2 + R2OH  + 2HCHO + C5H10Z       4.7000e12 0.0 -358.0 ! STUDY ANDERLOHR
R120C7H14OOOOH + NO => NO2 + R2OH  + 2HCHO + C5H10Z       4.7000e12 0.0 -358.0 ! STUDY ANDERLOHR
R121C7H14OOOOH + NO => NO2 + R2OH  + 2HCHO + C5H10Z       4.7000e12 0.0 -358.0 ! STUDY ANDERLOHR
R123C7H14OOOOH + NO => NO2 + R2OH  + 2HCHO + C5H10Z       4.7000e12 0.0 -358.0 ! STUDY ANDERLOHR
R126C7H14OOOOH + NO => NO2 + R2OH  + 2HCHO + C5H10Z       4.7000e12 0.0 -358.0 ! STUDY ANDERLOHR
R151C7H14OOOOH + NO => NO2 + R2OH  + 2HCHO + C5H10Z       4.7000e12 0.0 -358.0 ! STUDY ANDERLOHR
R155C7H14OOOOH + NO => NO2 + R2OH  + 2HCHO + C5H10Z       4.7000e12 0.0 -358.0 ! STUDY ANDERLOHR
R156C7H14OOOOH + NO => NO2 + R2OH  + 2HCHO + C5H10Z       4.7000e12 0.0 -358.0 ! STUDY ANDERLOHR
R157C7H14OOOOH + NO => NO2 + R2OH  + 2HCHO + C5H10Z       4.7000e12 0.0 -358.0 ! STUDY ANDERLOHR
R163C7H14OOOOH + NO => NO2 + R2OH  + 2HCHO + C5H10Z       4.7000e12 0.0 -358.0 ! STUDY ANDERLOHR
R164C7H14OOOOH + NO => NO2 + R2OH  + 2HCHO + C5H10Z       4.7000e12 0.0 -358.0 ! STUDY ANDERLOHR
R171C7H14OOOOH + NO => NO2 + R2OH  + 2HCHO + C5H10Z       4.7000e12 0.0 -358.0 ! STUDY ANDERLOHR
```

```
R128C8H16OOOOH + NO => NO2 + R2OH + 2HCHO + C6H12Z    4.7000e12 0.0 -358.0 ! STUDY ANDERLOHR
R129C8H16OOOOH + NO => NO2 + R2OH + 2HCHO + C6H12Z    4.7000e12 0.0 -358.0 ! STUDY ANDERLOHR
R130C8H16OOOOH + NO => NO2 + R2OH + 2HCHO + C6H12Z    4.7000e12 0.0 -358.0 ! STUDY ANDERLOHR
R133C8H16OOOOH + NO => NO2 + R2OH + 2HCHO + C6H12Z    4.7000e12 0.0 -358.0 ! STUDY ANDERLOHR
R136C8H16OOOOH + NO => NO2 + R2OH + 2HCHO + C6H12Z    4.7000e12 0.0 -358.0 ! STUDY ANDERLOHR
R137C8H16OOOOH + NO => NO2 + R2OH + 2HCHO + C6H12Z    4.7000e12 0.0 -358.0 ! STUDY ANDERLOHR
R138C8H16OOOOH + NO => NO2 + R2OH + 2HCHO + C6H12Z    4.7000e12 0.0 -358.0 ! STUDY ANDERLOHR
R139C8H16OOOOH + NO => NO2 + R2OH + 2HCHO + C6H12Z    4.7000e12 0.0 -358.0 ! STUDY ANDERLOHR
R141C8H16OOOOH + NO => NO2 + R2OH + 2HCHO + C6H12Z    4.7000e12 0.0 -358.0 ! STUDY ANDERLOHR
R142C8H16OOOOH + NO => NO2 + R2OH + 2HCHO + C6H12Z    4.7000e12 0.0 -358.0 ! STUDY ANDERLOHR
R143C8H16OOOOH + NO => NO2 + R2OH + 2HCHO + C6H12Z    4.7000e12 0.0 -358.0 ! STUDY ANDERLOHR
R145C8H16OOOOH + NO => NO2 + R2OH + 2HCHO + C6H12Z    4.7000e12 0.0 -358.0 ! STUDY ANDERLOHR
R146C8H16OOOOH + NO => NO2 + R2OH + 2HCHO + C6H12Z    4.7000e12 0.0 -358.0 ! STUDY ANDERLOHR
R148C8H16OOOOH + NO => NO2 + R2OH + 2HCHO + C6H12Z    4.7000e12 0.0 -358.0 ! STUDY ANDERLOHR
R176C8H16OOOOH + NO => NO2 + R2OH + 2HCHO + C6H12Z    4.7000e12 0.0 -358.0 ! STUDY ANDERLOHR
R180C8H16OOOOH + NO => NO2 + R2OH + 2HCHO + C6H12Z    4.7000e12 0.0 -358.0 ! STUDY ANDERLOHR
```

! DECOMPOSITIONS OF RO RADICALS BY BETA-SCISSION
```
R1C3H7O  => HCHO + R11C2H5        2.0e13 0.0 15000 ! publi Curran - valeur identique qqsoit position
R1C3H7O  => R1H + C2H5CHO         2.0e13 0.0 15000 ! publi Curran - valeur identique qqsoit position
R2C3H7O  => R4CH3 + CH3CHO        4.0e13 0.0 15000 ! publi Curran - valeur identique qqsoit position
R1C4H9O  => R1H + C3H7CHO         2.0e13 0.0 15000 ! publi Curran - valeur identique qqsoit position
R1C4H9O  => HCHO + R19C3H7        2.0e13 0.0 15000 ! publi Curran - valeur identique qqsoit position
R2C4H9O  => R1H + C3H7CHO         2.0e13 0.0 15000 ! publi Curran - valeur identique qqsoit position
R2C4H9O  => HCHO + R188C3H7       2.0e13 0.0 15000 ! publi Curran - valeur identique qqsoit position
R3C4H9O  => R4CH3 + C2H6CO        2.0e13 0.0 15000 ! publi Curran - valeur identique qqsoit position
R4C4H9O  => C2H5CHO + R4CH3       2.0e13 0.0 15000 ! publi Curran - valeur identique qqsoit position
R4C4H9O  => CH3CHO + R11C2H5      2.0e13 0.0 15000 ! publi Curran - valeur identique qqsoit position
R1C5H11O => R1H + C4H9CHO         2.0e13 0.0 15000 ! publi Curran - valeur identique qqsoit position
R1C5H11O => HCHO + R20C4H9        2.0e13 0.0 15000 ! publi Curran - valeur identique qqsoit position
R2C5H11O => R1H + C4H9CHO         2.0e13 0.0 15000 ! publi Curran - valeur identique qqsoit position
R2C5H11O => HCHO + R41C4H9        2.0e13 0.0 15000 ! publi Curran - valeur identique qqsoit position
R3C5H11O => R19C3H7 + CH3CHO      2.0e13 0.0 15000 ! publi Curran - valeur identique qqsoit position
R3C5H11O => C3H7CHO + R4CH3       2.0e13 0.0 15000 ! publi Curran - valeur identique qqsoit position
R4C5H11O => C2H5CHO + R11C2H5     4.0e13 0.0 15000 ! publi Curran - valeur identique qqsoit position
R1C7H15O => C3H7CHO + R19C3H7     2.0e13 0.0 15000 ! publi Curran - valeur identique qqsoit position
R2C7H15O => HCHO + R195C6H13      2.0e13 0.0 15000 ! publi Curran - valeur identique qqsoit position
R2C7H15O => R1H + C6H13CHO        2.0e13 0.0 15000 ! publi Curran - valeur identique qqsoit position
R3C7H15O => R4CH3 + C5H11CHO      2.0e13 0.0 15000 ! publi Curran - valeur identique qqsoit position
R3C7H15O => R33C5H11 + CH3CHO     2.0e13 0.0 15000 ! publi Curran - valeur identique qqsoit position
R4C7H15O => C4H9CHO + R11C2H5     2.0e13 0.0 15000 ! publi Curran - valeur identique qqsoit position
R4C7H15O => R20C4H9 + C2H5CHO     2.0e13 0.0 15000 ! publi Curran - valeur identique qqsoit position
R1C8H17O => R1H + C7H15CHO        2.0e13 0.0 15000 ! publi Curran - valeur identique qqsoit position
R1C8H17O => HCHO + R196C7H15      2.0e13 0.0 15000 ! publi Curran - valeur identique qqsoit position
R2C8H17O => R1H + C7H15CHO        2.0e13 0.0 15000 ! publi Curran - valeur identique qqsoit position
R2C8H17O => HCHO + R197C7H15      2.0e13 0.0 15000 ! publi Curran - valeur identique qqsoit position
R3C8H17O => R4CH3 + C6H14CO       2.0e13 0.0 15000 ! publi Curran - valeur identique qqsoit position
R3C8H17O => C2H6CO + R39C5H11     2.0e13 0.0 15000 ! publi Curran - valeur identique qqsoit position
R4C8H17O => C3H7CHO + R37C4H9     2.0e13 0.0 15000 ! publi Curran - valeur identique qqsoit position
R4C8H17O => C3H7CHO + R41C4H9     2.0e13 0.0 15000 ! publi Curran - valeur identique qqsoit position
```

! COUPLING OF ALDEHYDES WITH NO2
```
CH3CHO+NO2=>R4CH3+B2CO+HONO       8.35E-11 6.68 8300! XU05
C2H6CO+NO2=>R11C2H5+B2CO+HONO     8.35E-11 6.68 8300! XU05
C3H7CHO+NO2=>R19C3H7+B2CO+HONO    8.35E-11 6.68 8300! XU05
C3H7CHO+NO2=>R188C3H7+B2CO+HONO   8.35E-11 6.68 8300! XU05
C4H9CHO+NO2=>R20C4H9+B2CO+HONO    8.35E-11 6.68 8300! XU05
C4H9CHO+NO2=>R37C4H9+B2CO+HONO    8.35E-11 6.68 8300! XU05
C4H9CHO+NO2=>R41C4H9+B2CO+HONO    8.35E-11 6.68 8300! XU05
C4H9CHO+NO2=>R190C4H9+B2CO+HONO   8.35E-11 6.68 8300! XU05
C5H11CHO+NO2=>R33C5H11+B2CO+HONO  8.35E-11 6.68 8300! XU05
C5H11CHO+NO2=>R39C5H11+B2CO+HONO  8.35E-11 6.68 8300! XU05
C5H11CHO+NO2=>R56C5H11+B2CO+HONO  8.35E-11 6.68 8300! XU05
C5H11CHO+NO2=>R191C5H11+B2CO+HONO 8.35E-11 6.68 8300! XU05
C6H13CHO+NO2=>R195C6H13+B2CO+HONO 8.35E-11 6.68 8300! XU05
C6H13CHO+NO2=>R198C6H13+B2CO+HONO 8.35E-11 6.68 8300! XU05
C6H13CHO+NO2=>R199C6H13+B2CO+HONO 8.35E-11 6.68 8300! XU05
C7H15CHO+NO2=>R21C7H15+B2CO+HONO  8.35E-11 6.68 8300! XU05
C7H15CHO+NO2=>R22C7H15+B2CO+HONO  8.35E-11 6.68 8300! XU05
C7H15CHO+NO2=>R23C7H15+B2CO+HONO  8.35E-11 6.68 8300! XU05
C7H15CHO+NO2=>R24C7H15+B2CO+HONO  8.35E-11 6.68 8300! XU05
C7H15CHO+NO2=>R197C7H15+B2CO+HONO 8.35E-11 6.68 8300! XU05
C7H15CHO+NO2=>R196C7H15+B2CO+HONO 8.35E-11 6.68 8300! XU05
```

```
C7H15CHO+NO2=>R200C7H15+B2CO+HONO        8.35E-11  6.68  8300! XU05
C7H15CHO+NO2=>R201C7H15+B2CO+HONO        8.35E-11  6.68  8300! XU05
C7H15CHO+NO2=>R202C7H15+B2CO+HONO        8.35E-11  6.68  8300! XU05
```

!THE REACTIONS R.+NO2=composé
!Fall off Parameter Fc =0.183
```
C2H5NO2(+M)=R11C2H5+NO2(+M) 1.8E+17     0.0    58500.0
   LOw /1.3E+18 0.0 42000./
C3H7NO2(+M)=R19C3H7+NO2(+M) 1.8E+17     0.0    58500.0
   LOw /1.3E+18 0.0 42000./
C3H7NO2(+M)=R188C3H7+NO2(+M) 1.8E+17    0.0    58500.0
   LOw /1.3E+18 0.0 42000./
C4H9NO2(+M)=R20C4H9+NO2(+M) 1.8E+17     0.0    58500.0
   LOw /1.3E+18 0.0 42000./
C4H9NO2(+M)=R37C4H9+NO2(+M) 1.8E+17     0.0    58500.0
   LOw /1.3E+18 0.0 42000./
C4H9NO2(+M)=R41C4H9+NO2(+M) 1.8E+17     0.0    58500.0
   LOw /1.3E+18 0.0 42000./
C4H9NO2(+M)=R190C4H9+NO2(+M) 1.8E+17    0.0    58500.0
   LOw /1.3E+18 0.0 42000./
C5H11NO2(+M)=R33C5H11+NO2(+M) 1.8E+17   0.0    58500.0
   LOw /1.3E+18 0.0 42000./
C5H11NO2(+M)=R39C5H11+NO2(+M) 1.8E+17   0.0    58500.0
   LOw /1.3E+18 0.0 42000./
C5H11NO2(+M)=R56C5H11+NO2(+M) 1.8E+17   0.0    58500.0
   LOw /1.3E+18 0.0 42000./
C5H11NO2(+M)=R191C5H11+NO2(+M) 1.8E+17  0.0    58500.0
   LOw /1.3E+18 0.0 42000./
C7H15NO2(+M)=R21C7H15+NO2(+M) 1.8E+17   0.0    58500.0
   LOw /1.3E+18 0.0 42000./
C7H15NO2(+M)=R22C7H15+NO2(+M) 1.8E+17   0.0    58500.0
   LOw /1.3E+18 0.0 42000./
C7H15NO2(+M)=R23C7H15+NO2(+M) 1.8E+17   0.0    58500.0
   LOw /1.3E+18 0.0 42000./
C7H15NO2(+M)=R24C7H15+NO2(+M) 1.8E+17   0.0    58500.0
   LOw /1.3E+18 0.0 42000./
C7H15NO2(+M)=R196C7H15+NO2(+M) 1.8E+17  0.0    58500.0
   LOw /1.3E+18 0.0 42000./
C7H15NO2(+M)=R197C7H15+NO2(+M) 1.8E+17  0.0    58500.0
   LOw /1.3E+18 0.0 42000./
C7H15NO2(+M)=R200C7H15+NO2(+M) 1.8E+17  0.0    58500.0
   LOw /1.3E+18 0.0 42000./
C7H15NO2(+M)=R201C7H15+NO2(+M) 1.8E+17  0.0    58500.0
   LOw /1.3E+18 0.0 42000./
C7H15NO2(+M)=R202C7H15+NO2(+M) 1.8E+17  0.0    58500.0
   LOw /1.3E+18 0.0 42000./
C8H17NO2(+M)=R25C8H17+NO2(+M) 1.8E+17   0.0    58500.0
   LOw /1.3E+18 0.0 42000./
C8H17NO2(+M)=R26C8H17+NO2(+M) 1.8E+17   0.0    58500.0
   LOw /1.3E+18 0.0 42000./
C8H17NO2(+M)=R27C8H17+NO2(+M) 1.8E+17   0.0    58500.0
   LOw /1.3E+18 0.0 42000./
C8H17NO2(+M)=R28C8H17+NO2(+M) 1.8E+17   0.0    58500.0
   LOw /1.3E+18 0.0 42000./
```

!DECOMPOSITION REACTIONS : Y + NO2
```
RC4H7Y+NO2=> R4CH3+C2H3CHOZ+NO      2.35E+13  0.0   0.0 ! comme SLAgLE 1981
RC5H9Y+NO2=> R11C2H5+C2H3CHOZ+NO    2.35E+13  0.0   0.0 ! comme SLAgLE 1981
RC6H11Y+NO2=> R19C3H7+C2H3CHOZ+NO   2.35E+13  0.0   0.0 ! comme SLAgLE 1981
RC7H13Y+NO2=> R20C4H9+C2H3CHOZ+NO   2.35E+13  0.0   0.0 ! comme SLAgLE 1981
RC8H15Y+NO2=> R33C5H11+C2H3CHOZ+NO  2.35E+13  0.0   0.0 ! comme SLAgLE 1981
```

! REACTIONS : R.+NO2 => NO + RO.
```
R19C3H7 + NO2    => NO + R1C3H7O     4.0E+13  -0.2   0.0 !Glarborg 1999 pour Ch3+NO2=CH3O+NO
R188C3H7 + NO2   => NO + R2C3H7O     4.0E+13  -0.2   0.0 !Glarborg 1999 pour Ch3+NO2=CH3O+NO
R20C4H9 + NO2    => NO + R1C4H9O     4.0E+13  -0.2   0.0 !Glarborg 1999 pour Ch3+NO2=CH3O+NO
R41C4H9 + NO2    => NO + R2C4H9O     4.0E+13  -0.2   0.0 !Glarborg 1999 pour Ch3+NO2=CH3O+NO
R190C4H9 + NO2   => NO + R3C4H9O     4.0E+13  -0.2   0.0 !Glarborg 1999 pour Ch3+NO2=CH3O+NO
R37C4H9 + NO2    => NO + R4C4H9O     4.0E+13  -0.2   0.0 !Glarborg 1999 pour Ch3+NO2=CH3O+NO
R33C5H11 + NO2   => NO + R1C5H11O    4.0E+13  -0.2   0.0 !Glarborg 1999 pour Ch3+NO2=CH3O+NO
R39C5H11 + NO2   => NO + R2C5H11O    4.0E+13  -0.2   0.0 !Glarborg 1999 pour Ch3+NO2=CH3O+NO
R56C5H11 + NO2   => NO + R3C5H11O    4.0E+13  -0.2   0.0 !Glarborg 1999 pour Ch3+NO2=CH3O+NO
R191C5H11 + NO2  => NO + R4C5H11O    4.0E+13  -0.2   0.0 !Glarborg 1999 pour Ch3+NO2=CH3O+NO
```

```
!R195C6H13 + NO2  => NO + R1C6H13O    4.0E+13   -0.2   0.0 !Glarborg 1999 pour Ch3+NO2=CH3O+NO
!R198C6H13 + NO2  => NO + R1C6H13O    4.0E+13   -0.2   0.0 !Glarborg 1999 pour Ch3+NO2=CH3O+NO
!R199C6H13 + NO2  => NO + R1C6H13O    4.0E+13   -0.2   0.0 !Glarborg 1999 pour Ch3+NO2=CH3O+NO
R21C7H15 + NO2    => NO + R1C7H15O    4.0E+13   -0.2   0.0 !Glarborg 1999 pour Ch3+NO2=CH3O+NO
R22C7H15 + NO2    => NO + R2C7H15O    4.0E+13   -0.2   0.0 !Glarborg 1999 pour Ch3+NO2=CH3O+NO
R23C7H15 + NO2    => NO + R3C7H15O    4.0E+13   -0.2   0.0 !Glarborg 1999 pour Ch3+NO2=CH3O+NO
R24C7H15 + NO2    => NO + R4C7H15O    4.0E+13   -0.2   0.0 !Glarborg 1999 pour Ch3+NO2=CH3O+NO
R25C8H17 + NO2    => NO + R1C8H17O    4.0E+13   -0.2   0.0 !Glarborg 1999 pour Ch3+NO2=CH3O+NO
R26C8H17 + NO2    => NO + R2C8H17O    4.0E+13   -0.2   0.0 !Glarborg 1999 pour Ch3+NO2=CH3O+NO
R27C8H17 + NO2    => NO + R3C8H17O    4.0E+13   -0.2   0.0 !Glarborg 1999 pour Ch3+NO2=CH3O+NO
R28C8H17 + NO2    => NO + R4C8H17O    4.0E+13   -0.2   0.0 !Glarborg 1999 pour Ch3+NO2=CH3O+NO

!REACTIONS : R.+HONO=> NO2 + RH
!Abstraction of primary H:                    2.2E+13   0.0   31100 !CHAN2001
!Abstraction of secondary H:     5.8E+12      0.0   28100 !CHAN2001
!Abstraction of tertiary H:      9.3E+13      0.0   25800 !CHAN2001
C2H6+NO2=R11C2H5+HONO            2.2E+13      0.0   31100 !CHAN2001
C3H8+NO2=R19C3H7+HONO            2.2E+13      0.0   31100 !CHAN2001
C3H8+NO2=R188C3H7+HONO          5.8E+12      0.0   28100 !CHAN2001
C4H10+NO2=R20C4H9+HONO                      5.8E+12   0.0   28100 !CHAN2001
C4H10+NO2=R37C4H9+HONO           2.2E+13      0.0   31100 !CHAN2001
C4H10+NO2=R41C4H9+HONO           9.3E+13      0.0   25800 !CHAN2001
C4H10+NO2=R190C4H9+HONO          2.2E+13      0.0   31100 !CHAN2001
C5H12+NO2=R33C5H11+HONO          2.2E+13      0.0   31100 !CHAN2001
C5H12+NO2=R39C5H11+HONO          2.2E+13      0.0   31100 !CHAN2001
C5H12+NO2=R56C5H11+HONO          5.8E+12      0.0   28100 !CHAN2001
C5H12+NO2=R191C5H11+HONO         5.8E+12      0.0   28100 !CHAN2001
C6H14+NO2=R195C6H13+HONO         2.2E+13      0.0   31100 !CHAN2001
C6H14+NO2=R198C6H13+HONO         5.8E+12      0.0   28100 !CHAN2001
C6H14+NO2=R199C6H13+HONO         5.8E+12      0.0   28100 !CHAN2001
C7H16-1+NO2=R21C7H15+HONO        5.8E+12      0.0   28100 !CHAN2001
C7H16-1+NO2=R22C7H15+HONO        2.2E+13      0.0   31100 !CHAN2001
C7H16-1+NO2=R23C7H15+HONO        5.8E+12      0.0   28100 !CHAN2001
C7H16-1+NO2=R24C7H15+HONO        5.8E+12      0.0   28100 !CHAN2001
C7H16-1+NO2=R197C7H15+HONO       9.3E+13      0.0   25800 !CHAN2001
C7H16-1+NO2=R196C7H15+HONO       5.8E+12      0.0   28100 !CHAN2001
C7H16-1+NO2=R200C7H15+HONO       2.2E+13      0.0   31100 !CHAN2001
C7H16-1+NO2=R201C7H15+HONO       2.2E+13      0.0   31100 !CHAN2001
C7H16-1+NO2=R202C7H15+HONO       2.2E+13      0.0   31100 !CHAN2001
C8H18-1+NO2=R25C8H17+HONO        2.2E+13      0.0   31100 !CHAN2001
C8H18-1+NO2=R26C8H17+HONO        2.2E+13      0.0   31100 !CHAN2001
C8H18-1+NO2=R27C8H17+HONO        9.3E+13      0.0   25800 !CHAN2001
C8H18-1+NO2=R28C8H17+HONO        5.8E+12      0.0   28100 !CHAN2001

!REACTIONS : R.+HNO=> NO + RH
!cinetique comme CHOI2005 pour HNO+R4CH3=CH4+NO
!R4CH3+HNO    = NO+CH4    1.47E11  0.76  349 !CHOI2005 pour HNO+R4CH3=CH4+NO
R11C2H5+HNO   = NO+C2H6   1.47E11  0.76  349 !CHOI2005 pour HNO+R4CH3=CH4+NO
R19C3H7+HNO   = NO+C3H8   1.47E11  0.76  349 !CHOI2005 pour HNO+R4CH3=CH4+NO
R188C3H7+HNO  = NO+C3H8   1.47E11  0.76  349 !CHOI2005 pour HNO+R4CH3=CH4+NO
R20C4H9+HNO   = NO+C4H10  1.47E11  0.76  349 !CHOI2005 pour HNO+R4CH3=CH4+NO
R41C4H9+HNO   = NO+C4H10  1.47E11  0.76  349 !CHOI2005 pour HNO+R4CH3=CH4+NO
R190C4H9+HNO  = NO+C4H10  1.47E11  0.76  349 !CHOI2005 pour HNO+R4CH3=CH4+NO
R37C4H9+HNO   = NO+C4H10  1.47E11  0.76  349 !CHOI2005 pour HNO+R4CH3=CH4+NO
R39C5H11+HNO  = NO+C5H12  1.47E11  0.76  349 !CHOI2005 pour HNO+R4CH3=CH4+NO
R33C5H11+HNO  = NO+C5H12  1.47E11  0.76  349 !CHOI2005 pour HNO+R4CH3=CH4+NO
R56C5H11+HNO  = NO+C5H12  1.47E11  0.76  349 !CHOI2005 pour HNO+R4CH3=CH4+NO
R191C5H11+HNO = NO+C5H12  1.47E11  0.76  349 !CHOI2005 pour HNO+R4CH3=CH4+NO
R195C6H13+HNO = NO+C6H14  1.47E11  0.76  349 !CHOI2005 pour HNO+R4CH3=CH4+NO
R198C6H13+HNO = NO+C6H14  1.47E11  0.76  349 !CHOI2005 pour HNO+R4CH3=CH4+NO
R199C6H13+HNO = NO+C6H14  1.47E11  0.76  349 !CHOI2005 pour HNO+R4CH3=CH4+NO
R21C7H15+HNO  = NO+C7H16-1  1.47E11  0.76  349 !CHOI2005 pour HNO+R4CH3=CH4+NO
R22C7H15+HNO  = NO+C7H16-1  1.47E11  0.76  349 !CHOI2005 pour HNO+R4CH3=CH4+NO
R23C7H15+HNO  = NO+C7H16-1  1.47E11  0.76  349 !CHOI2005 pour HNO+R4CH3=CH4+NO
R24C7H15+HNO  = NO+C7H16-1  1.47E11  0.76  349 !CHOI2005 pour HNO+R4CH3=CH4+NO
R197C7H15+HNO = NO+C7H16-1  1.47E11  0.76  349 !CHOI2005 pour HNO+R4CH3=CH4+NO
R196C7H15+HNO = NO+C7H16-1  1.47E11  0.76  349 !CHOI2005 pour HNO+R4CH3=CH4+NO
R200C7H15+HNO = NO+C7H16-1  1.47E11  0.76  349 !CHOI2005 pour HNO+R4CH3=CH4+NO
R201C7H15+HNO => NO+C7H16-1  1.47E11  0.76  349 !CHOI2005 pour HNO+R4CH3=CH4+NO
R202C7H15+HNO = NO+C7H16-1  1.47E11  0.76  349 !CHOI2005 pour HNO+R4CH3=CH4+NO
R25C8H17+HNO  = NO+C8H18-1  1.47E11  0.76  349 !CHOI2005 pour HNO+R4CH3=CH4+NO
R26C8H17+HNO  = NO+C8H18-1  1.47E11  0.76  349 !CHOI2005 pour HNO+R4CH3=CH4+NO
R27C8H17+HNO  = NO+C8H18-1  1.47E11  0.76  349 !CHOI2005 pour HNO+R4CH3=CH4+NO
```

```
R28C8H17+HNO  = NO+C8H18-1   1.47E11  0.76  349 !CHOI2005 pour HNO+R4CH3=CH4+NO

!***********************************************************************
!*                                                            *
!*          REACTION BASE : TOULENE - NOX                     *
!*                                                            *
!***********************************************************************

!REACTIONS OF BENZENE AND PHENYL RADICALS
C6H6#+NO2=C6H5#+HONO          7.4E13   0.0   38200   !CHAN2001
C6H6#+NO2=C6H5#+HNO2          2.5E14   0.0   42200   !CHAN2001
C6H5NO2(+M)=C6H5#+NO2(+M)     1.52E17  0.0   73717   !XU2005
       LOw /1.52E17   0.0     73717/
HNO+C6H5#=C6H6#+NO            3.78E5   2.28  456     !CHOI2005
HNO+C6H5#=C6H5NO+R1H          3.79E9   1.19  95400   !CHOI2005
C6H5NO(+M)=C6H5#+NO(+M)       1.52E17  0.0   55200   !TSENG2004
       LOw /1.52E17   0.0     55200/

!REACTIONS OF PHENYL RADICALS
C6H5O2+NO=C6H5O#+NO2          4.7E12   0.0   -358.0   !Estimated comme ROO+NO

!REACTIONS OF TOLUENE AND BENZYL RADICALS
benzyl+HNO2=NO2+toluene           8.1E+4   1.87  4838
benzyl+HONO=NO2+toluene           8.1E+4   1.87  5504
HNO+benzyl=toluene+NO             1.47E10  0.76  349
HNO+C6H4CH3=toluene+NO            3.78E5   2.28  456
benzyl+NO2=C6H5CH2O+NO            1.36E12  0.0   0
C6H5CH2NO2(+M)=benzyl+NO2(+M) 1.8E17  0.0  58500.0
       LOw /1.3E+18 0.0 42000./
!             SRI /1.0 0.0 1.E-18 0.183 0.0/

!REACTIONS OF BENZYLPEROXY RADICALS
C6H5CH2OO+NO=NO2+C6H5CH2O     4.70E12  0.0   -358   !Estimated comme ROO+NO

!REACTIONS OF BENZYLALKOXY RADICALS
C6H5CH2O+NO=C6H5CHO+HNO       7.6E13   -0.76  0
C6H5CH2O+NO2=C6H5CHO+HONO     4.0E12   0.0   2285
HNO+C6H5CH2O=C6H5CH2OH+NO     3.16E13  0.0   0

!REACTIONS OF BENZALDEHYDE RADICALS
C6H5CHO+NO2=C6H5CO+HONO       8.35E-10  6.68  8300
C6H5CHO+NO=C6H5CO+HNO         1.02E13   0.0  40670

!REACTIONS OF HOC6H4CH2 RADICALS
HOC6H4CH2+HNO=HOC6H4CH3+NO    1.47E11  0.76  349
HOC6H4CH2+NO2=HOC6H4CH2O+NO   1.36E12  0.0   0
HOC6H4CH2+HONO=HOC6H4CH3+NO2  8.1E+4   1.87  5504

!REACTIONS OF C6H5CH2NO2
C6H5CH2NO2+R2OH=C6H5CHO+NO+H2O   3.0E6          2.0   2000
C6H5CH2NO2+B1O=C6H5CHO+NO+R2OH   1.51E13  0.0   5354
C6H5CH2NO2+R1H=C6H5CHO+NO+H2     4.67E12  0.0   3732
C6H5CH2NO2+R4CH3=C6H5CHO+NO+CH4  7.08E11  0.0   11140
C6H5CH2NO2+R4CH3=HCHO+NO+toluene 7.08E11  0.0   11140

!REACTIONS OF C6H5NO2
C6H5NO2(+M)=C6H5O#+NO(+M)     7.12E13  0.0   62590
C6H5NO+NO2=C6H5NO2+NO         9.62E10  0.0   12928   !PARK2002

END
```

A.2 Oxidation of PRF and toluene containing fuels in absence of NO_x

We tested our PRF/toluene/NO_x oxidation model for applications without NO_x interactions, as well. We present here validations performed against autoignition delay data obtained in rapid compression machines, shock tubes and HCCI engines. Further we provide validations for species profiles measured for slow oxidations in JSR under highly diluted conditions. Experimental conditions used for in JSR experiments can be found in Table 16. The summaries of the experimental conditions used for ST, RCM and HCCI facilities are presented in Table 32 and Table 33.

Table 32   *Summary of experimental conditions used for auto-ignition simulations in rapid compression machines (RCM) and shock tubes (ST).*

Author	Exp. setup	Fuel Reference	FUEL n-C_7 [mol/mol]	i-C_8 [mol/mol]	C_7H_8 [mol/mol]	Dilutant	$P_{reactor}$ [atm]	$T_{reaction}$ [K]	ϕ [-]
Vanhove et al. (2006)	RCM	Fuel_Van_1	1.0	0.0	0.0	CO_2, N_2, Ar	3.3-4.5	650-910	1.0
	RCM	Fuel_Van_2	0.0	1.0	0.0		12.4-15.8	660-880	1.0
	RCM	Fuel_Van_3	0.5	0	0.5		3.8-4.8	650-860	1.0
	RCM	Fuel_Van_4	0.0	0.65	0.35		12.-14.6	670-850	1.0
Callahan et al. (1996)	RCM	Fuel_Cal_1	0.0	1.0	0.0	CO_2, N_2, Ar	11-17	650-910	1.0
	RCM	Fuel_Cal_2	0.55	0.95	0.0		11–17	650-910	1.0
	RCM	Fuel_Cal_3	0.1	0.9	0.0		11-17	650-910	1.0
Tanaka et al. (2003)	RCM	Fuel_Tan_1	0.26	0.0	0.74	N_2	40-44	798-878	0.4
	RCM	Fuel_Tan_2	0.21	0.05	0.74		40-44	798-878	0.4
	RCM	Fuel_Tan_3	0.16	0.1	0.74		40-44	798-878	0.4
Ciezki et al. (1993)	ST	Fuel_Cie_1	1.0	0.0	0.0	CO_2, N_2	13	650-1200	0.5-2
Fieweger et al. (1997)	ST	Fuel_Fiew_1	0.0	1.0	0.0	CO_2, N_2	40	650-1200	1.0
	ST	Fuel_Fiew_2	0.1	0.9	0.0		40	650-1200	1.0
	ST	Fuel_Fiew_3	0.2	0.8	0.0		40	650-1200	1.0
	ST	Fuel_Fiew_4	0.4	0.6	0.0		40	650-1200	1.0
	ST	Fuel_Fiew_5	1.0	0.0	0.0		40	650-1200	1.0
Gauthier et al. (2004)	ST	Fuel_Gau_1	0.17	0.55	0.28	N_2	12 and 60	860 - 1022	1.0
	ST	Fuel_Gau_2	0.17	0.63	0.2		12 and 60	860 - 1022	1.0

Table 33 Summary of experimental conditions used for simulations in HCCI-engines.

Author	Exp. setup	Fuel Reference	FUEL n-C_7 $[mol/mol]$	FUEL Iso-C_8 $[mol/mol]$	FUEL C_7H_8 $[mol/mol]$	NO [ppm]	P_0 [atm]	T_0 [K]	ϕ [-]	Speed [rpm]	CR [-]
	HCCI	Fuel_Andr_1 (Run_1)	0.06	0.94	0.0	0	1	393	0.29	900	16.7
	HCCI	Fuel_Andr_1 (Run_2)	0.06	0.94	0.0	0	2	313	0.25	900	16.7
Andrae et al. (2005)	HCCI	Fuel_Andr_2 (Run_2)	0.16	0.84	0.0	0	2	313	0.25	900	16.7
	HCCI	Fuel_Andr_3 (Run_1)	0.25	0.0	0.75	0	1	393	0.29	900	16.7
	HCCI	Fuel_Andr_3 (Run_2)	0.25	0.0	0.75	0	2	313	0.25	900	16.7
	HCCI	Fuel_Andr_4 (Run_2)	0.35	0.0	0.65	0	2	313	0.25	900	16.7

A.2.1 Rapid Compression Machines (RCM) and Shock Tubes (ST)

Figure A-1 and Figure A-2 compare experimental and simulated results for rapid compression machine experiments performed by Vanhove et al. (2006). The authors measured both cool flame (t_{cf}) and main ignition (t_{ig}) delay times for pure n-heptane, pure iso-octane, n-heptane/toluene and iso-octane/toluene blends in a rapid compression machine as a function of temperature. They obtained temperature variations in a range from 600 to 900 K at the end of compression by varying the composition of the inert gas. The temperature obtained at the end of the compression was calculated using the adiabatic core gas model. The authors deduced ignition and cool flame delay times from the pressure and light-emission traces.

Their experimental results for the oxidation of pure n-heptane (Figure A-1a) and iso-octane (Figure A-1b) reveal that at temperatures around 800 K there is a zone with increasing ignition delays associated to increasing temperatures, which is characteristic of the NTC-regime. Compared to the delays measured for the oxidation of pure n-heptane and iso-octane, the addition of toluene to n-heptane (Figure A-2a) and to iso-octane (Figure A-2b) increases the ignition delays and reduces the NTC effect at intermediate temperatures.

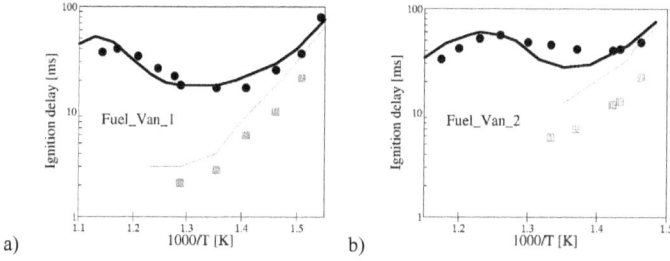

Figure A-1   *Experimental (symbols) and simulated (lines) ignition delay times of cool flame (green squares □ and thin full line -) and main flame- (black circles ● and thick full line —) ignition delays obtained in a **RCM** (Vanhove et al. 2006) for the stoichiometric oxidation of (a) **n-heptane** at pressures around **4 atm** and (b) **iso-octane** at pressures around **14 atm**. Exp*

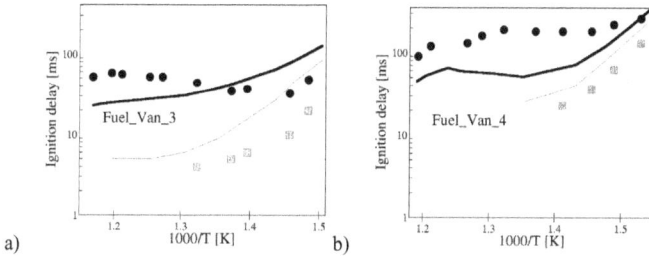

Figure A-2   *Experimental (symbols) and simulated (lines) ignition delay times of cool flame (green squares □ and thin full line -) and main flame (black circles ● and thick full line —) ignition delays obtained in a **RCM** (Vanhove et al. 2006) for the stoichiometric oxidation of (a) **n-heptane/toluene** mixtures at pressures around **4 atm** and (b) **iso-octane/toluene** mixtures at pressures around **14 atm**. Exp*

The simulations show that the ignition delays of both cool and main flames produced by our mechanism are qualitatively correct. However, in the case of pure iso-octane (Figure A-1b), cool flame delays and main ignition delays above 800 K are overestimated. The retarding impact of toluene on the n-heptane and iso-octane oxidation is retrieved (Figure A-2), showing stronger toluene interactions in the case of iso-octane. For the oxidation of toluene/n-heptane mixtures (Figure A-2a), the mechanism overestimates ignition delays below 700 K and underestimates them at above 800 K (Figure A-2a). Nevertheless, the appearance of cool flames and main flame ignition delays are qualitatively well captured. Compared to experiments, simulations predict generally shorter ignitions delays for the oxidation of iso-octane/toluene mixtures (Figure A-2b). This is believed to be caused by an insufficient coupling of iso-octane/toluene for capturing the retarding impact of toluene on the iso-octane oxidation.

Callahan et al. (1996) measured main flame ignition delays for n-heptane/iso-octane mixtures at pressures between 11 and 17 atm by varying the RON number, characterised by the iso-octane/n-heptane ratio. Figure A-3 displays measured and simulated main flame ignition delays for a RON of 100, 95 and 90. Experiments and simulations show decreasing ignition delays for decreasing RON. The simulations qualitatively retrieve ignition delays over the analyzed temperature range, including the NTC effect, but generally ignition delays are slightly overestimated. This is especially the case for PRF 90 where the accelerating impact of n-heptane on the iso-octane ignition is not well reproduced. This is similar to the results obtained for the oxidation of pure iso-octane presented in Figure A-1b.

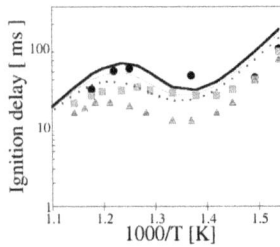

*Figure A-3 Experimental (symbols) and simulated (lines) main flame ignition delay times obtained in a **RCM** (Callahan et al. 1996) for the stoichiometric oxidation of PRF mixtures at pressures around **15 atm** with a **RON** of **100** (black circles ● and thick full line ━), **95** (green squares ▢ and thin full line -) and **90** (blue shaded triangles ⊿ and dotted lines --). Exp*

Ciezki et al. (1993) and Fieweger et al. (1997) measured ignition delay times with shock tube experiments at around 13 atm and varied the equivalence ratio for the oxidation of pure n-heptane. They defined ignition delays by visualizing CH emissions and interpreting pressure signals and obtained an accuracy of around ±20µs. Figure A-4 shows the comparison between measured data and simulated results. Experiments and simulations both show decreasing ignition delay times for increasing equivalence ratios. Generally, ignition delays for n-heptane are captured correctly, even if at low temperatures the model tends to overpredict ignition delays.

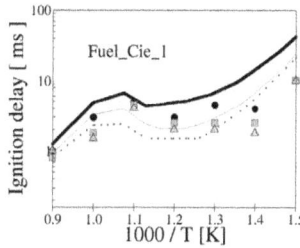

*Figure A-4   Experimental (symbols) and simulated (lines) ignition delay times obtained in a ST (Ciezki et al. 1993) for **n-heptane** air mixtures at **13 atm**, for an **equivalence ratio** of **0.5** (black circles ● and thick full line ▬), **1.0** (green squares ▫ and thin full line -) and **2.0** (blue triangles △ and dotted line ---). Exp*

Fieweger et al. (1997) performed shock tube experiments for the stoichiometric oxidation of PRF-mixtures at pressures around 40 atm. They varied the RON number of the investigated PRF mixtures from 0 to 100. Experimental data and simulation results are displayed in Figure A-5. Experiments show a significant impact of the RON number on ignition delays for a temperature range between 600 and 1200 K, which is reproduced well by the model.

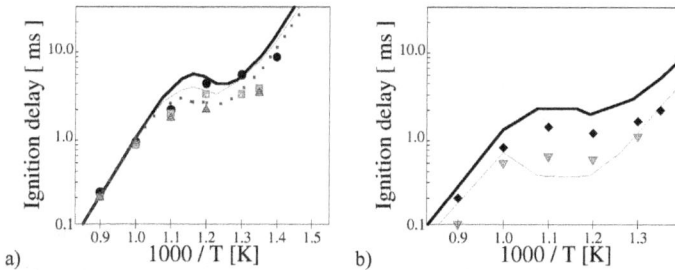

*Figure A-5   Experimental (symbols) and simulated (lines) ignition delay times obtained in a ST (Fieweger et al. 1997) at around **40 atm** for the stoichiometric oxidation of **PRF** air mixtures with a **RON** of (a) **100** (black circles ● and thick full line ▬), **90** (green squares ▫ and thin full line -) **80** (blue triangles △ and dotted line ...) and (b) **60** (black diamonds ◆ and thick full line ▬) and **0** (red triangles △ and thin full line -).*

Tanaka et al. (2003a) have studied the autoignition behaviour of different fuels in a rapid compression machine. Their study focuses on the sensitivity of ignition delay times to small variations of the iso-octane/n-heptane ratio with a fixed amount of toluene. They obtained mixture purities greater than 99 %, recorded the pressure histories and deduced temperature evolutions by using isentropic relations. Fractional errors in the recorded peak pressures and ignition delay times are less than ±3 %. Recorded results are compared to simulations for n-heptane/iso-octane/toluene mixtures and are shown in Figure A-6.

Experimental results of all investigated fuel blends reveal the appearance of a cool flame, correctly reproduced by simulations. The stepwise increase of the iso-octane concentration causes a non-linear increase of the main ignition delay which is also captured.

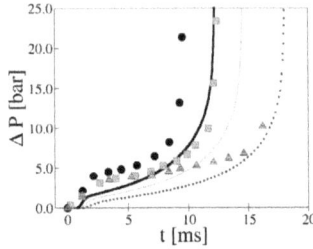

*Figure A-6   Experimental (symbols) and simulated (lines) pressure profiles obtained in a RCM (Tanaka et al. 2003a) at a pressure of **10 atm** for the oxidation of* **Fuel_Tan_1** *(black circles ● and thick full line ▬),* **Fuel_Tan_2** *(green squares □ and thin full line -) and* **Fuel_Tan_3** *(blue shaded triangles △ and dotted line ---) at an equivalence ratio of 0.4. Exp*

Gauthier et al. (2004) measured ignition delay times of ternary n-heptane/iso-octane/toluene mixtures in a ST. They performed experiments at medium (15-20 atm) and high pressures (45-60 atm), varying temperatures from 850 to 1100 K. In the considered temperature range the obtained discrepancies of recorded delay times are indicated by the authors as being less than 10 μs. Their study examines the sensitivity of ignition delay times by varying the toluene/iso-octane ratio with a constant n-heptane concentration. The comparison between experimental data and simulation results is shown in Figure A-7. Experiments show that a pressure increase reduces the ignition delays. At high pressures, ignition delays become quasi independent of the toluene/iso-octane ratio (compare Figure A-7a and Figure A-7b), which is not the case at lower pressures. In the pressure range from 15 to 20 atm and for an increased toluene/iso-octane ratio retarded ignition delays are observed. The described characteristics are well reproduced by the model. However an overprediction of ignition delays is generally observed in the investigated temperature and pressure ranges.

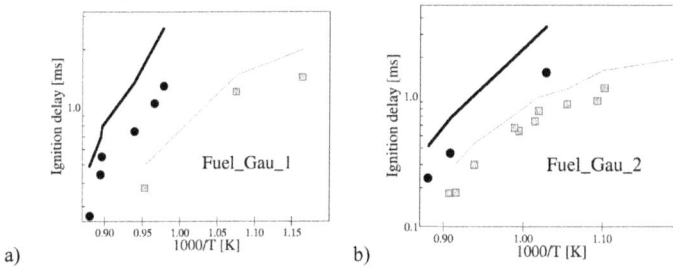

*Figure A-7 Experimental (symbols) and simulated (lines) main flame ignition delay times obtained in a **ST** (Gauthier et al. 2004) for the stoichiometric oxidation of (a) **Fuel_Gau_1** at intermediate pressures (12-25 atm) (black circles ● and thick line —) and at high pressures (45-60 atm) (green squares □ and thin line -) and (b) of **Fuel_Gau_2** at intermediate pressures (12-25 atm) (black circles ● and thick line —) , at high pressures (45-60 atm) (green squares □ and thin line -). Exp*

A.2.2 HCCI engines

Validations in HCCI engines are crucial because they allow the model to be tested on realistic configurations. Andrae et al. (2007) have performed experiments in an HCCI-engine with PRF and n-heptane/toluene mixtures. They tested 2 different experimental set-ups varying equivalence ratios, initial engine temperatures and initial engine pressures. The engine was operated at 900 rpm and the pressure in the HCCI-engine was measured as a function of Crank Angle Degree (CAD) After Top Dead Center (ATDC). Figure 8a shows experimental and simulated results obtained for PRF mixtures characterized by two different RON. Two different engine runs (Run_1 and Run_2 in Table 33) were tested varying admission temperature and equivalence ratio. Experimental and simulated data are shown for Fuel_Andr_1 (6 % n-heptane, 94 % iso-octane) for the engine operation conditions Run_1 and Run_2 and compared to the engine combustion of Fuel_Andr_2 for engine operation condition Run_2. Figure A-8b presents the analogous comparison between experimental and simulated data for the two different toluene/n-heptane mixtures (Fuel_Andr_3, Fuel_Andr_4).

Experimental results show that lowering the iso-octane/n-heptane ratio (Figure A-8a) and decreasing the toluene/n-heptane-ratio (Figure A-8b) results in an increased reactivity and a faster fuel ignition. This behaviour is well captured by the model. In addition, the impact of pressure and temperature variations leading to a faster fuel ignition for the engine set-up Run_2 is also correctly predicted. The overprediction of maximum pressures is due to the simplifying assumption of an adiabatic combustion and a perfectly stirred fuel/air mixture.

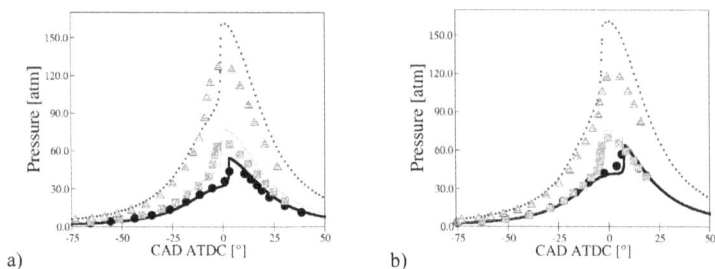

*Figure A-8   Experimental (symbols) and simulated (lines) pressure profiles obtained in a **test engine**
(Andrae et al. 2007) for (a) engine Run_1 with **Fuel_Andr_1** (black circles ● and thick
full line ▬), engine Run_1 with **Fuel_Andr_2** (green squares □ and thin lines -), engine
Run_2 with **Fuel_Andr_2** (blue shaded triangles △ and dotted lines ---) and (b) engine
Run_1 with **Fuel_Andr_3** (black circles ● and thick full line ▬), engine Run_1 with
Fuel_Andr_4 (green squares □ and thin full line -), engine Run_2 with **Fuel_Andr_4**
(blue shaded triangles △ and dotted line ---).*

A.2.3 Jet Stirred Reactors (JSR)

Validations in JSR are of importance because they allow the model to be tested for the
consumption and formation of intermediate species. Moréac et al. (2003) studied the
oxidation of n-heptane, iso-octane and toluene at different pressures and for temperatures
from 500 to 1100 K. They recorded reactant mole fractions (including NO and NO_2) by using
chromatographic techniques with uncertainties of ±10 % except for toluene and benzene,
where the uncertainties reached up to ±15 %. Figure A-9 and Figure A-10 compare
experimental and simulated results for the oxidation of n-heptane at 1 atm and 10 atm,
respectively. At atmospheric pressure no n-heptane conversion (Figure A-9a) and no
production of CO (Figure A-9b) is observed below 900 K. Simulations satisfactorily
reproduce the non-reactivity of n-heptane at low temperatures. At 10 atm pure n-heptane
starts reacting at temperatures around 520 K (Figure A-10a) associated to a CO production
(Figure A-10b) which is correctly represented by the model. However, simulations
underestimate the strong NTC behaviour observed experimentally between 700 and 800 K
shown in experiments.

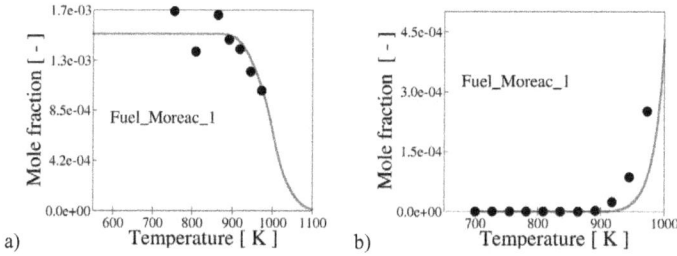

Figure A-9   *Experimental (symbols) and simulated (lines) concentration profiles of (a) n-heptane mole fractions and (b) CO mole fractions obtained in a **JSR** (Moréac et al. 2003) for the stoichiometric oxidation of 1500 ppm of **n-heptane (Fuel_Moreac_1)** at a pressure of 1 **atm**.*

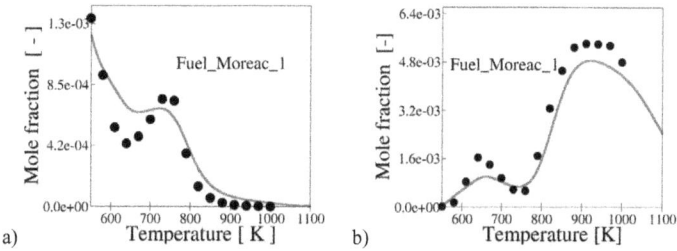

Figure A-10   *Experimental (symbols) and simulated (lines) concentration profiles of (a) n-heptane mole fractions and (b) CO mole fractions obtained in a **JSR** (Moréac et al. 2003) for the stoichiometric oxidation of 1500 ppm of **n-heptane (Fuel_Moreac_1)** at a pressure of **10 atm**.*

Figure A-11 compares experimental data of Moréac et al. (2003) to simulated results for the stoichiometric oxidation of iso-octane at atmospheric pressure. As for n-heptane, no reactant conversion (Figure A-11a) and no CO production (Figure A-11b) are observed below 900 K in experiments and simulations.

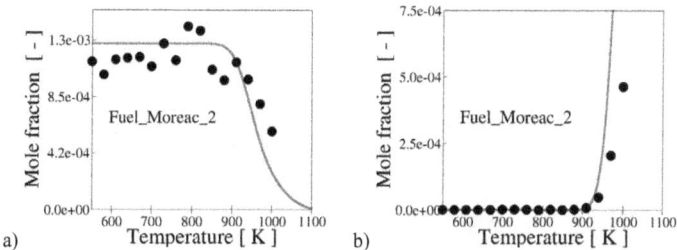

Figure A-11   *Experimental (symbols) and simulated (lines) concentration profiles of (a) iso-octane mole fractions and (b) CO mole fractions obtained in a **JSR** (Moréac et al. 2003) for the stoichiometric oxidation of 1250 ppm of **iso-octane (Fuel_Moreac_2)** at a pressure of **1 atm**.*

Experimental results of Moréac et al. (2003) and simulations for the oxidation of toluene at a pressure of 10 atm are shown in Figure A-12. Experiments reveal that toluene starts reacting at temperatures around 900 K. This is observed by a strong diminution of its concentration (Figure A-12a) and a strong increase of that of CO (Figure A-12b). These characteristics are correctly captured by the model.

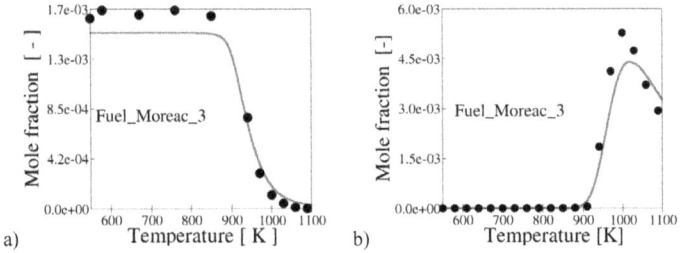

a)   b)

Figure A-12 Experimental (symbols) and simulated (lines) concentration profiles of (a) toluene mole fractions and (b) CO mole fractions obtained in a **JSR** (Moréac et al. 2003) for the stoichiometric oxidation of 1500 ppm of **toluene** (**Fuel_Moreac_3**) at a pressure of **10 atm**.

Dubreuil et al.(2007) performed experiments in a JSR for the oxidation of PRF mixtures and for n-heptane/toluene mixtures at a pressure of 10 atm. The authors indicate a fuel purity greater than 99.9 %, while they obtained variations in the carbon balance of measured species of around ±5 %. The comparison of measured and simulated iso-octane concentrations presented in Figure A-13 also shows good agreement with experimental observations.

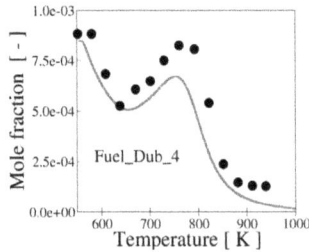

Figure A-13 Experimental (symbols) and simulated (lines) concentration profiles of iso-octane mole fractions obtained in a **JSR** (Dubreuil et al. 2007) for the lean oxidation (ϕ=0.2) of **Fuel_Dub_4** at a pressure of **10 atm** for an addition of 0 ppm NO.

The experimental and computed data for the oxidation of n-heptane/toluene mixtures are shown in Figure A-14. The temperature at which the n-heptane/toluene mixture reacts is around 580 K which corresponds to the value measured by Moréac et al. (2003) for the oxidation of pure n-heptane. Under similar conditions, pure toluene only starts to react around

900 K (Moréac et al. 2003) indicating the promoting effect caused by the presence of n-heptane on the toluene oxidation. A good agreement between simulation and experimental results is obtained.

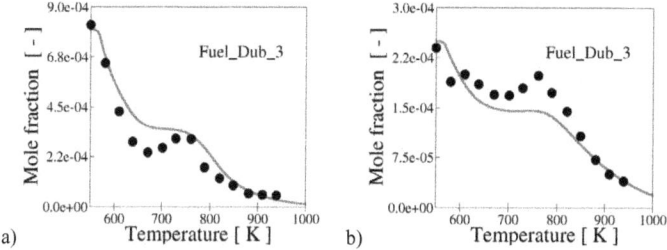

*Figure A-14   Experimental (symbols) and simulated (lines) concentration profiles of (a) n-heptane mole fractions and (b) toluene mole fractions obtained in a **JSR** (Dubreuil et al. 2007) for the lean oxidation (ϕ=0.2) of **Fuel_Dub_3** at a pressure of **10 atm**.*

Appendix B

FPI-Tabulation methodology

B.1 Definition of the progress variable

Different definitions of the progress variable are proposed in literature (Fiorina et al. (2003), Colin et al. (2004, 2005), Ribert et al. (2006) Embouazza et al. (2003), Mauviot et al. (2006)). Our initial intention was a definition of the progress variable similar to that proposed by Mauviot et al. 2006, who defines c by a notation of y_c as follows:

$$[48] \qquad y_c = \left(y_{O_2}^0 - y_{O_2}\right) + y_{CO} + y_{CO_2}$$

This definition allows an excellent reproduction of the cool-flame phenomena. Under cool flame conditions, oxygen is consumed despite limited production of CO and CO_2. For that reason the definition of c chosen by Mauviot et al. (2006) is more accurate in the cool flame region than the definition of c by CO, CO_2 fractions only. However, the definition of c proposed by Mauviot et al. (2006) depends on the initial oxygen concentration y_{O2}^0 at the start of reaction when c equals 0. The numerical precision of the Chemkin software used for generating the FPI-table and the numerical precision of the $IFPC3D$-code is limited. For that reason, the initial concentration of oxygen at reaction start $c=0$ in the CFD-code does not correspond exactly to the initial concentration of oxygen for which detailed chemistry computations were performed and which are stored in the FPI-table. This incoherence generates a small numerical error in the reaction advancement c, which is integrated over the complete computation time. An important divergence of the c-evolution in time is the consequence resulting in great errors in the temporal reproduction of chemical kinetics. Such problem could not be observed by Mauviot et al. (2006). Their strategy was the tabulation of species-reaction rates instead of species-mass fractions. In that case, the composition error in initial oxygen is not integrated over the reaction time. In that case reaction rates are tabulated and not reconstructed from the species composition. The described error in c remains negligible, because it is not integrated over the reaction time. The described problem of integrated errors in the c-evolution in time let us chose the commonly used definition of c by CO and CO_2 as it was defined in equations [16] and [17].

$$[16] \qquad y_c = y_{CO} + y_{CO_2}$$

$$[17] \qquad c = \frac{y_c - y_c^0}{y_c^{eq} - y_c^0}$$

At reaction start ($c=0$) the concentrations of the combustion products CO and CO_2 equal exactly 0. That is true for the $IFPC3D$-code as well as for the Chemkin computations by which the FPI-table is generated. Thus, no numerical errors in the progress variable occur and the correct evolution of c in time is ensured.

B.2 Progress of species mass fractions y_i in time t

Figure B-1 illustrates a reaction Manifold (red line) which is discretized (vertical dotted lines) over the reaction time t. y_i denotes for the species mass- fraction obtained by the CFD-computations and c_j^{FPI} represent the tabulated reaction advancements of the tabulated manifold. t is the current time, dt the current time step and τ is a characteristic time scale allowing to relax towards tabulated chemistry.

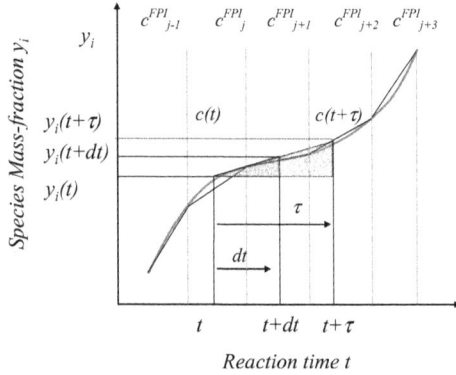

Figure B-1   Advancement in time of species mass fractions along a FPI-manifold.

A reactive mixture should be at the state defined by advancement $c(t)$ and it should be advanced towards $c(t+dt)$. At the next lower tabulated point $c_{(j-1)}^{FPI}$ the reaction rate ω_{j-1}^{FPI} is read. ω_{j-1}^{FPI} is stored in the FPI-table and was calculated from detailed chemistry by,

$$[49] \qquad \omega_{j-1}^{FPI} = \frac{c_j^{FPI} - c_{j-1}^{FPI}}{t_j - t_{j-1}}$$

where $c_{(j)}^{FPI}$ denotes for the chosen c-values for the discretization of each manifold in c. The reaction advancement at instant $c(t+\tau)$ is calculated as follows:

$$[21]a \qquad c(t + \tau) = c(t) + \tau \cdot \omega_{j-1}^{FPI}$$

The mass-fractions y_i $(t+\tau)$ corresponding to the reaction advancement $c(t+\tau)$ are linearly interpolated from the tabulated species mass fractions at the upper and lower tabulated neighbour points in c, y_{j+2}^{FPI} and y_{j+1}^{FPI} as shown in equation [23].

$$[50] \qquad y_i(t + \tau) = y_{i,j+1}^{FPI} + \frac{y_{i,j+2}^{FPI} - y_{i,j+1}^{FPI}}{c_{j+2}^{FPI} - c_{j+1}^{FPI}} \cdot \left(c(t+\tau) - c_{j+1}^{FPI} \right)$$

The mass-fraction $y_i(t+dt)$ are finally obtained by the linear integration of the time gradient of $y_i(t+\tau)$ and $y_i(t)$ over dt:

$$[23] \qquad y_i(t+dt) = y_i(t) + \frac{y_i(t+\tau) - y_i(t)}{\tau} \cdot dt$$

B.3 Sensitivity of kinetics to potential table characteristics

The main consequence of the limitations in table size and dimensions is that not all potential thermochemical characteristics of postoxidation may be tabulated. Thus, a first challenge is to evaluate the impact of each variable representing one table dimension on reaction kinetics and to check if it might be neglected. Figure B-2 to Figure B-9 show the ignition delays as they were computed for the stoichiometric oxidation of a mixture consisting of 13.7 mol% of n-heptane, 42.9 mol% of iso-octane and 43.4 mol% of toluene diluted by 97% of N_2 (i.e. the same mixture composition as for simulations of section 5.1). We have tested the impact of variations of the thermochemical characteristics temperature (T_0) (Figure B-2), pressure (P_0) (Figure B-3), equivalence ratio (y_{CxHy_0}, y_{O2_0}, SAI_0) (Figure B-4), CO-mass fractions (y_{CO_0}) (Figure B-5), CO_2-mass fractions (y_{CO2_0}) (Figure B-6), H_2O-mass fractions (y_{H2O_0}) (Figure B-7) and NO-mass fractions (y_{NO_0}) (Figure B-8) on ignition delays. In the postoxidation case, the equivalence ratio is defined by the fuel mass fraction (y_{CxHy_0}), oxygen mass fraction (y_{O2_0}) and fraction of secondary air (SAI_0) injected into the exhaust gases. Temperature evolutions are plotted over the non-dimensional time τ. τ was normalized by the ignition delay obtained for the stoichimetric mixture oxidation at 1 bar, a temperature of 800 K and oxidized with a dilution by pure N_2 (CO, CO_2, H_2O = 0 ppm) of 97 mol%.

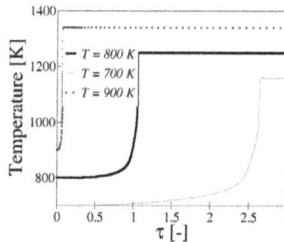

Figure B-2 Temperature evolution obtained for the stoichiometric oxidation of the chosen fuel surrogate at 97 mol% N_2 dilution, at atmospheric pressure and at different temperatures.

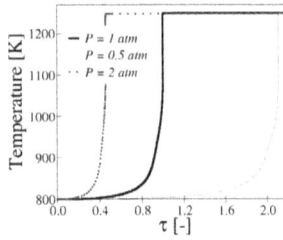

Figure B-3 Temperature evolution obtained for the stoichiometric oxidation of the chosen fuel surrogate at 97 mol% N_2 dilution, a temperature of 800 K and at different pressures.

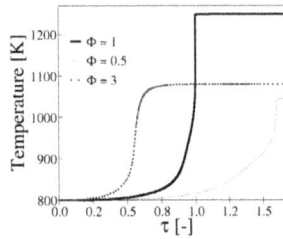

Figure B-4 Temperature evolution obtained for the oxidation of the chosen fuel surrogate at 97 mol% N_2 dilution at atmospheric pressure, a temperature of 800 K and at different equivalence ratios.

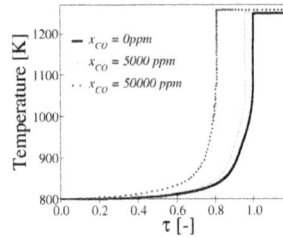

Figure B-5 Temperature evolution obtained for the stoichiometric oxidation of the chosen fuel surrogate at 97 mol% N_2 dilution at atmospheric pressure, a temperature of 800 K and at different initial concentrations of CO.

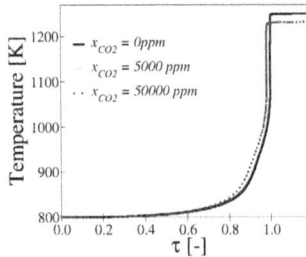

Figure B-6 Temperature evolution obtained for the stoichiometric oxidation of the chosen fuel surrogate at 97 mol% N_2 dilution at atmospheric pressure, an initial temperature of 800 K and at different initial concentrations of CO_2.

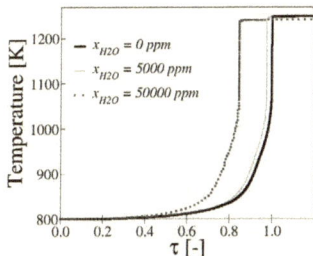

Figure B-7 Temperature evolution obtained for the stoichiometric oxidation of the chosen fuel surrogate at 97 mol% N_2 dilution at atmospheric pressure, an initial temperature of 800 K and at different initial concentrations of H_2O.

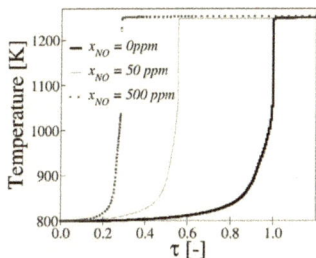

Figure B-8 Temperature evolution obtained for the stoichiometric oxidation of the chosen fuel surrogate at 97 mol% N_2 dilution at atmospheric pressure, an initial temperature of 800 K and at different initial concentrations of NO.

Our analysis reveals that, as expected, chemical kinetics are extremely sensitive to temperature variations (Figure B-2). A temperature reduction or increase of ±100 K changes ignition delays by a factor of up to 5. The pressure under the investigated conditions is another determining property for reaction kinetics too (Figure B-3). Doubling or dividing the atmospheric pressure by two results in an increase or a reduction of ignition delays by a factor of around 2 or 0.5 respectively. Kinetics are rather sensitive to variations in equivalence ratio (Figure B-4). Changing the equivalence ratio between 0.5 and 3 leads to a reduction/increase of ignition delays of around 50-60% compared to the stoichiometric mixture. Generally, ignition delays of the n-heptane/iso-octane/toluene mixture decrease with increasing equivalence ratio.

The presence of CO shows an accelerating impact on reaction kinetics at the tested conditions (Figure B-5). In presence of 5 mol% of CO, the ignition delay is by 20% shortened than for the corresponding mixture in the absence of CO. CO_2 in comparison shows a very slight impact on reactivity (Figure B-6). Mixture reactivity is also sensitive to the presence of H_2O (Figure B-7) very similar to the influence of CO (Figure B-5). The presence of 5 mol%

of H_2O results in a reduction of the ignition delay by around 20% compared to the corresponding mixture without H_2O. Chemical kinetics are very sensitive to the presence of NO (Figure B-8). NO present in an order of magnitude of less than 100 ppm may change ignition delay times by a factor 0.5. This analysis reveals that all of the investigated thermochemical characteristics must be considered for reproducing exhaust gas oxidation kinetics correctly. Neither the thermodynamic conditions, temperature or pressure, nor the mixture composition of the exhaust gases in terms of equivalence ratio or dilutant composition (N_2, CO, CO_2 and H_2O) can be neglected.

B.4 Dilutant composition at reaction equilibrium

We tested the impact of pressure, temperature and equivalence ratio on the equilibrium composition of PRF/toluene/air mixtures and performed CHEMKIN equilibrium calculations. Figure B-9 shows the mass fractions of CO, CO_2 and H_2O at reaction equilibrium as a function of equivalence ratio ϕ for various pressures P and temperatures T_{eq}.

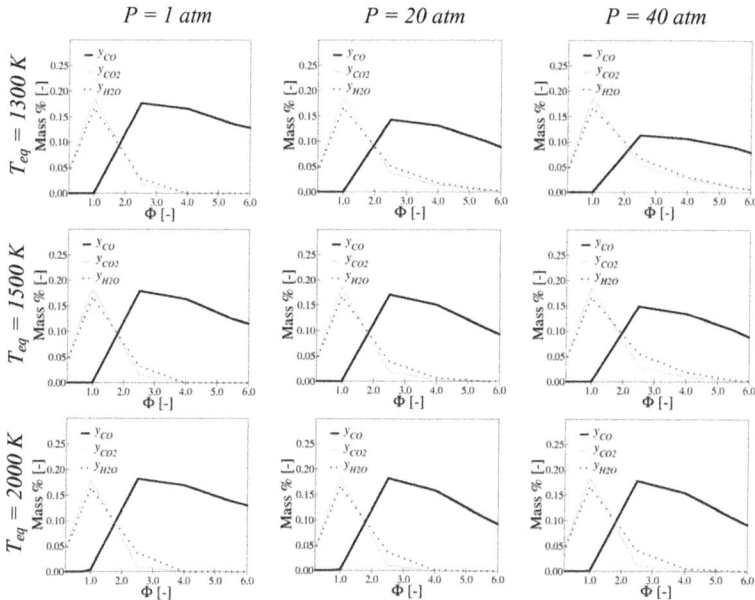

Figure B-9 *Mass fractions of dilutant components at reaction equilibrium computed under constant pressure conditions for a pure PRF/toluene/air ($y_{EGR}=0$) mixtures at various equilibrium temperatures T_{eq} and pressures P.*

Figure B-10 reveals that the temperature T_{eq} has only a slight impact on the residual gas composition (CO, CO_2 and H_2O). However, CO mass fractions are slightly reduced under rich conditions ($\phi > 2$) when the pressure P increases. This might be due to favoured recombination of CO with unburned hydrocarbons at increased pressures.

B.5 Table generation and properties of the new FPI-table

This section describes the generation of the FPI-table following the principles described above. All detailed chemistry computations were performed with the Chemkin II software package to solve a closed homogenous reactor at constant pressure. An automatic table generation tool has been developed to manage the large number of chemical simulations necessary to build the FPI look-up table.

B.5.1 Detailed chemistry computations

A basic principle of the FPI methodology is that the initial thermochemical conditions of the CFD-computations must be consistent with the thermochemical conditions under which the detailed chemistry calculations were performed when the FPI-table was generated. Hence, the FPI-table was created by analogous routines similar to those built for the reconstruction of the composition space at postoxidation computation start. The species mass fractions defined as input data for the detailed chemistry computations were imposed from the following set of equations.

$$[40] \qquad y_{Fuel}^{init} = \frac{q_{BG}/W_{BG}}{q_{Fuel}/W_{Fuel}}$$

$$[41] \qquad y_{CO_2}^{init} = f(y_{Fuel}^{init})$$

($y_{CO_2}^{init}$ assumed to be 0 for the co-flow case described in chapter 8)

$$[42] \qquad y_{H_2O}^{init} = f(y_{Fuel}^{init})$$

$$[43] \qquad y_{CO}^{init} = 0$$

$$[44] \qquad y_{O_2}^{init} = y_{O_2}^{IFPC3D}$$

$$[45] \qquad y_{NO}^{init} = y_{NO}^{IFPC3D}$$

$$[46] \qquad y_{N_2}^{init} = 1 - y_{Fuel}^{init} - y_{O_2}^{init} - y_{H_2O}^{init} - y_{CO_2}^{init} - y_{NO}^{init}$$

where y_i^{init} represent the mass fractions of species i (Fuel, CO, CO_2, H_2O, N_2, NO) computed in absence of injected secondary air. When secondary air is present the mass fractions of total O_2 and N_2 are computed as follows.

$$[51] \qquad y_{O_2-tot}^{init} = y_{O_2}^{IFPC3D} + \alpha \cdot \frac{0.23}{0.77}$$

$$[52] \qquad y_{N_2-tot}^{init} = y_{N_2}^{IFPC3D} + \alpha \cdot \frac{0.77}{0.23}$$

Thus the total mass-fraction y_{tot} can be computed by equation [53].

$$[53] \quad y_{tot} = y_{Fuel}^{init} + y_{O_2\_tot}^{init} + y_{N_2\_tot}^{init} + y_{CO}^{init} + y_{CO_2}^{init} + y_{H_2O}^{init} + y_{NO}^{init}$$

When secondary air is present y_{tot} exceeds 1 and thus species mass-fractions have to be renormalized by equation [55].

$$[54] \quad y_i^{Chemkin} = \frac{y_i^{init}}{y_{tot}}; \text{ for } y_i^{init} = y_{Fuel}^{init}, y_{O_2\_tot}^{init}, y_{N_2\_tot}^{init}, y_{CO}^{init}, y_{CO_2}^{init}, y_{H_2O}^{init}, y_{NO}^{init};$$

The conversion from mass to molar fractions is defined as:

$$[55] \quad x_i^{Chemkin} = \frac{\dfrac{y_i^{Chemkin}}{M_i}}{\sum \dfrac{y_i^{Chemkin}}{M_i}}$$

where M_i represents the molecular weight of species i, $y_i^{Chemkin}$ the mass fractions of tabulated species i and $x_i^{Chemkin}$ the molar fractions of tabulated species i. After detailed chemistry computations, the results which are given by Chemkin in molar fractions must be re-converted to mass fractions before storage in the FPI-table.

B.5.2 Definition of reaction equilibrium

The FPI-methodology is based on the principle that all reactive system tends to approach its thermodynamic equilibrium. Hence, the definition of the progress variable c is based on normalization over species mass fractions at their thermodynamic equilibrium and the correct definition of the thermodynamic equilibrium is therefore important. Pera et al. (2009) have defined the reaction equilibrium by computing each reactive system up to times scales close to 'infinity' $(t > 1 \times 10^5 \text{s})$. Figure B-10 shows, in logarithmic scale, the temperature evolutions of a reactive system at an equivalent ratio ϕ of 6. The temperature evolution is plotted over time in logarithmic scale.

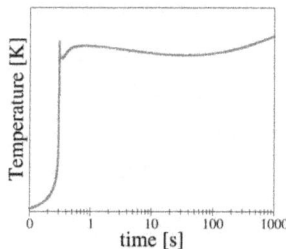

Figure B-10   Temperature evolution of a reactive mixture at an equivalence ratio of $\phi = 6$ computed up to very large time scales close to 'infinity'.

After ignition, a local temperature maximum is achieved but immediately after, a drop-off occurs in a time scale comparable to that of ignition. With increasing time, slow heat release is observed again and even after a time $t > 10^3$ s, the mixture continues to react and the temperature is still rising. Such behaviour has no physical meaning but it is an artefact of detailed kinetic reaction mechanisms when dealing with conditions beyond the ones for which they have been thoroughly validated. This is the case for very high equivalence ratios. Detailed reaction mechanisms are built for predicting species evolutions during oxidation and auto-ignition, but not for predicting thermodynamic equilibrium in those conditions. In some detailed reaction mechanisms, reactions written in the forward direction only can be found.

$$[56] \qquad A + B \to C$$

Such reactions are not reversible and hence, they do not respect thermodynamic equilibrium. A non physical accumulation of species may occur and in the case of exothermic reactions, a temperature rise up to very large time scales is observed. Generally reaction products C minor species. However, under very rich conditions, they can accumulate. For that reason, thermodynamic equilibrium is defined here as the state at which the accumulated CO and CO_2 mass-fractions reach a local maximum value. Figure B-11 indicates the defined reaction equilibrium and illustrates for the same computation shown in Figure B-10 the time evolution of the accumulated CO and CO_2 mass fractions as they are stored in the FPI-table.

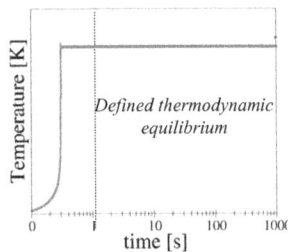

Figure B-11   Evolution of $y_{CO}+y_{CO2}$ fraction of a reactive mixture at an equivalence ratio of $\phi = 6$ computed up to a very large time close to 'infinity'.

We defined the reaction equilibrium at a reaction time of 1 s which physically will never be attempted during postoxidation computations in IC-engines. A comparison between residual gas composition at reaction equilibrium defined here and the equilibrium calculated based on only thermodynamic data (CHEMKIN: Equilibrium computations) showed acceptable agreement.

B.5.3 Numerical stability of detailed chemistry computations

It was mentioned that for generating our table 30.720 computations of detailed chemistry have to be performed. Hence, numerical stability of the applied computations is of major importance for an automatization of the table generation. The observed numerical divergences can be classified into two types:

- Numerical failures induced by strong gradients
- Numerical failures induced by weak gradients

Figure B-12 and Figure B-13 illustrate these two kind of potential numerical computation failures.

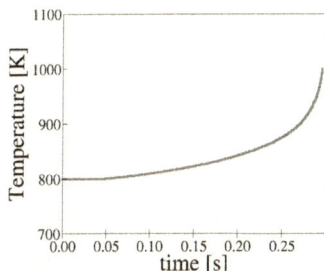

Figure B-12   Numerical computational failure at a temperature of 1000 K induced by a strong temperature gradient.

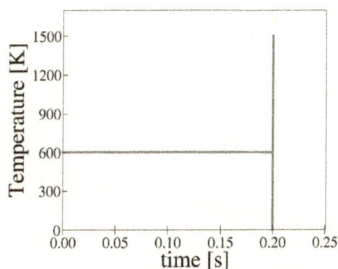

Figure B-13   Numerical computational failure at low temperature oxidation (T = 600 K) induced by weak gradients.

Numerical stiffness of Chemkin calculations is controlled via the definition of an absolute tolerance (ATOL) and a relative tolerance (RTOL). The absolute tolerance is used by the solvers as an indicator of the accuracy desired in the physical solution. The relative tolerance is used by the solver to determine convergence and indicates the accuracy desired in the physical solution. Generally the value of RTOL corresponds roughly to the number of significant digits that should be expected from a solution. We performed a parameter study on the ATOL and RTOL parameters observing numerical robustness.

Numerical failures resulting from strong gradients were observed, when the ATOL, RTOL parameters were chosen too small (strict tolerance). At strong gradients the solver may not obtain the limits of strictly defined solution tolerance and thus fails (see Figure B-12). Numerical failures from weak gradients were observed, when the ATOL, RTOL parameters were chosen too big (weak tolerance). In that case small gradients in the solution are not captured any more by the solver and 0-gradients are computed, without integration of small gradients, which could not be captured. Such conditions may also result in numerical computation failure as shown in Figure B-13. Best numerical stiffness was observed for an ATOL/RTOL parameter set of:

$$[57] \quad ATOL = 1 \cdot 10^{-20} \ or \ 1 \cdot 10^{-19}$$

$$[58] \quad RTOL = 1 \cdot 10^{-8} \ or \ 1 \cdot 10^{-7}$$

Beside the parameters determining numerical robustness we detected another major cause of computation failure at strong gradients. We observed that nearly all failures induced by strong gradients occur at 1000 K. For that reason we checked thermodynamic species data defined in the detailed reaction mechanism. Thermodynamic species data are defined according to the NASA-code which defines the heat capacity C_p, enthalpy H and entropy S of each species as a function of temperature by 2 sets of polynomials. The first defines the thermodynamic properties at high temperatures, the second at low temperatures. For most species the shift from one polynomial set to another occurs at 1000 K. In all tested reaction mechanisms we observed strong discontinuities in the C_p, H and S at the temperatures at which the parameter set is shifted. Figure B-14 illustrates for a representative species a discontinuity in the C_p – temperature evolution.

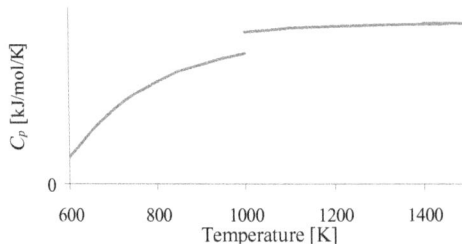

Figure B-14   Evolution of the calorific power C_p of a representative species as it is calculated by the polynomial coefficient defined in the NASA-code showing the typical discontinuity at 1000 K.

Such discontinuities cause major numerical convergence problems at 1000 K. We then developed a tool which shifts the C_p, enthalpy H and entropy S temperature evolutions defined for the high temperature range onto the levels defined for the low temperature region. Discontinuities could be eliminated and the numerical robustness was increased drastically.

B.6 Reconstruction of the composition space by element balances

The species reconstruction by molar and element balances is based on the principle of mass conservation as C, O and H elements are balanced by three reconstructed species. In IFP-C3D, nine transported species mass fractions are necessary during IC-engine computations with the FPI model of Pera et al. (2009). These are:

- y_{Fuel},	- y_{CO}	- y_{CO2}
- y_{H2O},	- y_{H}	- y_{N2}
- y_{O2}	- y_{H2}	- y_{C7H14}.

Among these, five (y_{Fuel}, y_{CO}, y_{CO2}, y_{H2O} and y_{H}) are tabulated and consequently, their concentrations correspond to the kinetics stored in the FPI-table. N_2 is considered as an inert dilutant. It is transported but does not participate at any combustion process. In the Pera et al. (2009) model implemented in IFP-C3D, O_2, H_2 and C_7H_{14} are reconstructed species. Hence, their computed concentrations contain no information about molecular oxygen, molecular hydrogen and heptene which are physically present in the burned gases. These reconstructed species only indicate the quantity of accumulated partially oxidized compounds not transformed from fuel into the tabulated intermediates and products CO, CO_2, •H and H_2O. There is no information about the real physical species which they represent.

The most important information necessary for a correct initialization of postoxidation computations is the fuel concentration and the diluting species present at the start of the process. CO, CO_2 and H_2O at the end of IC-engine oxidations are well defined because they are tabulated. However, dealing with fuel left over in the burned gases is slightly more complicated. Fuel C_xH_y is a tabulated species. However under rich conditions, it might on one hand be completely consumed, but on the other hand be partially oxidized or decomposed into intermediate hydrocarbons without leading to CO and CO_2 production. In that case, the tabulated fuel concentrations are close to 0 at the end of combustion, while the reconstructed species do not contain any information about the real composition they represent. Hence, the challenge is to correctly reconstruct the unburned fuel and partially oxidized compounds which may later react during postoxidation.

The idea of species reconstruction by molar- and element-balances might be to reconstruct partially oxidized hydrocarbons from element balances. This results in equation systems of different orders depending of the number of species that should be reconstructed. We tested different equation systems for reconstructing partially oxidized species with an order of up to 7 and faced major problems in solving numerically such systems. We finally abandoned the strategy of species reconstruction by molar- and element- balances, because such systems are stringently resolved in mol fractions. Nevertheless in *IFPC3D* species concentrations are stored in mass fractions and drastic errors are committed when mass fractions computed by *IFP-C3D* are converted to mol fractions. We thus rebuilt the composition space by alternative strategies described in chapter 7.4.2.

www.ingramcontent.com/pod-product-compliance
Lightning Source LLC
Chambersburg PA
CBHW021034210326
41598CB00016B/1015